LabVIEW for Automotive, Telecommunications, Semiconductor, Biomedical, and Other Applications

ISBN 0-13-019963-X

NATIONAL INSTRUMENTS | **VIRTUAL INSTRUMENTATION SERIES**

Lisa K. Wells • Jeffrey Travis
■ LabVIEW For Everyone

Mahesh L. Chugani • Abhay R. Samant • Michael Cerna
■ LabVIEW Signal Processing

Barry Paton
■ Sensors, Transducers & LabVIEW

Rahman Jamal • Herbert Pichlik
■ LabVIEW Applications and Solutions

Jeffrey Travis
■ Internet Applications in LabVIEW

Shahid F. Khalid
■ LabVIEW Applications/CVI Programming For Beginners

Hall T. Martin • Meg L. Martin
■ LabVIEW For Automotive, Telecommunications, Semiconductor, Biomedical, and Other Applications

Jeffrey Y. Beyon
■ Hands-On Exercise Manual For LabVIEW Programming, Data Acquisition, and Analysis

Jeffrey Y. Beyon
■ LabVIEW Programming, Data Acquisition, and Analysis

LabVIEW for Automotive, Telecommunications, Semiconductor, Biomedical, and Other Applications

▲ Hall T. Martin
▲ Meg L. Martin

Prentice Hall PTR, Upper Saddle River, NJ 07458
www.phptr.com

Library of Congress Cataloging-in-Publication Data

Martin, Hall T.
  LabVIEW for automotive, telecommunications, semiconductor, biomedical, and other applications. /
    Hall T. Martin, Meg L. Martin
      p. cm.
  Includes index.
  ISBN 0-13-019963-X (alk. paper)
      1. Engineering instruments—data processing.   2. LabVIEW.   I. Martin, Meg L.   II. Title.

TA 165 .M35 2000
620.'0028--dc21
                                                                  00-039982
                                                                         CIP

Editorial/Production Supervision: Wil Mara
Acquisitions Editor: Bernard Goodwin
Editorial Assistant: Diane Spina
Marketing Manager: Lisa Konzelmann
Manufacturing Manager: Alexis R. Heydt
Cover Design: Talar Agasyan
Cover Design Direction: Jerry Votta
Series Design: Gail Cocker-Bogusz

© 2000 Prentice Hall PTR
Prentice-Hall, Inc.
Upper Saddle River, New Jersey 07458

Prentice Hall books are widely used by corporations and government agencies for training, marketing, and resale. The publisher offers discounts on this book when ordered in bulk quantities. For more information, contact Prentice Hall's Corporate Sales Department—phone: 1-800-382-3419; fax: 1-201-236-7141; email: corpsales@prenhall.com; address: Corp. Sales Dept., Prentice Hall PTR, 1 Lake Street, Upper Saddle River, NJ 07458. All rights reserved. No part of this book may be reproduced, in any form or by any means, without permission in writing from the publisher.

ComponentWorks™, CVI™, DAQCard™, FieldPoint™, FlexMotion™, HiQ™, IMAQ™, IVI™, LabVIEW™, National Instruments™, NI-DAQ™, NI-IMAQ™, PXI™, SCXI™, and TestStand™ are trademarks of National Instruments Corporation.

Product and company names mentioned herein are trademarks or trade names of
their respective companies.

Printed in the United States of America
10 9 8 7 6 5 4 3 2 1

ISBN 0-13-019963-X

Prentice-Hall International (UK) Limited, *London*
Prentice-Hall of Australia Pty. Limited, *Sydney*
Prentice-Hall Canada Inc., *Toronto*
Prentice-Hall Hispanoamericana, S.A., *Mexico*
Prentice-Hall of India Private Limited, *New Delhi*
Prentice-Hall of Japan, Inc., *Tokyo*
Pearson Education Asia P.T.E., Ltd.
Editora Prentice-Hall do Brasil, Ltda., *Rio de Janeiro*

Acknowledgements

We would like to thank the following people for their contributions to this project.

Special thanks to our steering committee, which read and judged papers for this book. External committee members included Behbood Zoghi, Ph.D., Texas A&M University; Bill McKinnon, Ph.D., Georgia Institute of Technology; Roy Craig, Ph.D., University of Texas at Austin; Vittal Prabhu, Ph.D., Penn State University; Hossein Mousavinezhad, Western Michigan University.

National Instruments steering committee members included Thierry Debelle, Don Holley, Leo Little, Prahalad Vasudev, and Dave Wilson.

In addition, thanks to Maggie Ingram and Amber Turner for all their help in editing and preparing the application papers for publication.

Contents

Preface xviii

1
Automotive Test 1

Introduction 1
    Why is Automotive Important? 1
    What are the Present Trends and Challenges? 1
    What are the Future Trends and Challenges? 3
    How does National Instruments Fit In? 4
Electric Vehicle Inverter Durability Test Stand 5
    Introduction 5
    System Description 6
    Results 9

Machine Vision in Automotive Instrument Manufacturing 11
Introduction 11
Objective 12
System Description 13
Calibration Utility 16
Cluster Inspection Utility 19
Results 20

Tire Rolling Resistance Measurements Using LabVIEW and VXI 22
Introduction 22
Software 23
Hardware 26
Results 27

PC-Based Control of a Gasoline-Fueled Burner Aging Test Stand to Simulate Engine Exhaust 29
Introduction 29
System Design 30
Results 32

In-Vehicle Data Acquisition, Transfer, and Real-Time Processing 35
Introduction 35
Data Acquisition 36
Test Execution and Data Acquisition Synchronization 37
Real-Time Analysis 39
Review Test Data Utility 40
Results 41

Automotive Audio Test System 43
    Introduction 43
    Hardware Design Considerations 44
    Hardware Design 45
    Software Design Considerations 46
    Software Design – Administration Utility 47
    Software Design – Client Executable Application 47
    Design Challenges 50
    Results 51
Ever Take a Picture of a Pothole From a Moving Truck? 52
    Introduction 52
    Hardware Design 55
    Software Design 55
    Results 57
Automated Radio Tester 58
    Introduction 58
    Tests 60
    Test Parameters 61
    Test Results 62
    Test Scheduling 64
    Results 65

▼ 2
Biomedical Test 66

Introduction 66
   Why is Biomedical Important? 66
   What are the Present Trends and Challenges? 67
   What are the Future Trends and Challenges?........68
   How does National Instruments Fit In? 69
Cutting Latency on Assessing Heart Period Variability Studies 70
   Introduction..71
   System Hardware..72
   System Software...72
   Benchmark Measurements 74
   Results 75
A Cardiovascular Pressure-Dimension Analysis System 76
   Introduction...76
   The System..77
   Data Acquisition and Analysis 78
   Clinical Significance...82
   LabVIEW Tips and Techniques 84
   Conclusion 84
PC-Based Vision System for Wound Healing Assessment 87
   Introduction...87
   System Hardware and Software...........................89
   Results 89

Biomedical Patient Monitoring, Data Acquisition,
and Playback with LabVIEW 92
    Introduction 92
    Design Challenge and Solution 93
    Application 96
    Results 97

▼ 3
Semiconductor Test 99

Introduction 99
    Why is Semiconductor Test Important? 99
    What are the Present Trends? 100
    What are the Future Trends and Challenges? 102
    How Does National Instruments Fit In? 102
Angular Scanning Ellipsometer (ASELL) 103
    Introduction and System Requirements 103
    System Design 105
    Results 108
Graphical Modeling of Quantum Atomic State Transitions
in Hydrogenic Atoms 111
    Introduction 111
    Wave Functions Under an Applied Field 112

    Separation of Variables in Wave Functions 113
    Graphical Representation of State Transitions 114
    LabVIEW Implementation 115
    Some of the Plots Produced by LabVIEW 116
    Results 119
Intelligent Automation of Electron Beam Physical
Vapor Deposition 121
    Introduction 121
    Monitoring and Analysis 122
    Results 125
Control System for X-Ray Photolithography Tool 126
    Introduction 126
    Control System 127
    User Interface 128
    Conclusion 129
Data Acquisition From a Vacuummeter Controlled by
RS-232 Standard Using LabVIEW 131
    Introduction 131
    Front Panel Description 133
    Block Diagram Description 135
    Results 137

4
Telecommunications Test 138

Introduction 138
    What is Telecommunications? 138
    What are the Present Trends and Challenges? 140
    What is in the Future? 141
    How is National Instruments Involved? 141
LabVIEW-Based Antenna Measurements 142
    Introduction 142
    Multipath Distortion 143
    Hardware Control Software 143
    Results 145
Telecommunications Protocol Analysis Tool 147
    Introduction 147
    System Description 148
    Results 151
Remote Diagnostics in a Fiber Optic Network 152
    Introduction 152
    Program Design 153
    Results 155
Quick Real-Time Test of Communication Algorithm
Using LabVIEW 156
    System Hardware Control 156
    Benchmark of System Costs 157

System Cost Reduction 157
Development Time Reduction 158
Results 158
Common Test Software for Cellular Base Stations 160
Introduction 160
Requirements 161
Software Model 162
Estimated Return on Investment 166
Results 166

5
General Test 168

Introduction 168
Why is General Test Important? 168
What are the Present Trends and Challenges? 169
What are the Future Trends and Challenges? 171
How does National Instruments Fit In? 172
LabVIEW-Based Interactive Teaching Laboratory 173
Introduction 173
System Architecture 175
Sample User Interfaces 178
Remote Area Experiments over the Internet 181
Results 182

PC-Based Data Acquisition System for Measurement
and Control of an Isotopic Exchange Installation 185
    Introduction 186
    Product Development Performances 187
    Results 190
LabVIEW Tests M1A1 Ammunition 191
    Introduction 191
    Testing Procedure 192
Industrial X-Ray Digital Image Flaw Detection System 195
    Introduction 195
    System Hardware 196
    System Software 196
    System Function and Performance 197
    Results 199
Large Area Conditioning Systems for the National
Ignition Facility 200
    Introduction 200
    Development 201
    Beam Profiler 202
    Analog Data Acquisition 203
    User Interface and Execution Systems 204
    Results 206
Controlling Aeronautical Hydraulic Actuator Testing
with LabVIEW 208
    Introduction 208
    Custom SCXI Module 209

System Layout 211
　　Benefits 214
　　Results 215
Automating the San Francisco Bay Model with LabVIEW 217
　　Introduction 217
　　System Upgrade 218
　　Automation Requirements 219
　　System Configuration 219
　　Results 222
Virtual Balancing Equipment for Rigid Rotors 223
　　Introduction 223
　　System Hardware Configuration 224
　　Virtual Balancer Options 225
　　Two-Plane, Two Accelerometers Balancer 225
　　Innovative Features 228
　　Auxiliary Windows 229
　　Advantages of the Equipment 229
　　Results 231
LabVIEW-Based Automation of a Direct-Write Laser Beam
Micromachining System 232
　　Introduction 232
　　The Micromachining Application: Process Automation 234
　　System Requirements 234
　　Database Concepts: Expanding the Materials Space 236
　　Results 240

Future Improvements 240
Conclusions 241
Acknowledgements 241

Index 243

Preface

A cursory glance into a test lab 20 years ago likely revealed a nondescript computer connected by a standard cable to a bulky electronic instrument. The computer helped its owner write and print reports and maybe do some analysis, but traditional instruments, oscilloscopes, and multimeters did the brunt of the measurement work.

In those early years of computer-based measurement and automation, the desktop computer, linked by the General Purpose Interface Bus (GPIB), played an auxiliary role; however, the increasingly powerful PC has changed all of that. Today, the PC can acquire, analyze, and present data at increasing frequencies, resolutions, and sampling rates. PCs, which now perform tasks 1,000 times more quickly than their predecessors of just 10 years ago, can perform more specialized and complex work in vastly different industries. Engineers and scientists have capitalized on this increasing ability of modern PCs and the highly productive software and hardware that now runs on these machines to monitor everything from the human heart to computer chips to cell phones.

For more than 20 years, National Instruments™ has provided engineers and scientists with the hardware and software that has made the PC at home in the test lab and the manufacturing floor. With National Instruments tools, engineers now can place the computer at the center of complex measurement and automation systems, and by doing so, they have lowered their costs. Growth of standard computer technology, fueled by consumer demand in the 1980s and 1990s, increased productivity in the lab and on the manufacturing floor. Upgrading complex measurement equipment no longer meant purchasing expensive controllers and instruments dedicated to specific tasks. Instead, engineers and scientists could use desktop computers for thousands of dollars less than traditional equipment.

The equipment that National Instruments has offered for more than 20 years has come ready to interface with National Instruments flexible software. Data acquisition and signal conditioning devices, instrument control interfaces, image acquisition, motion control, and industrial communication devices make changing the type of measurement as simple as switching the hardware device connected to your computer.

Now, the National Instruments computer technology that has lowered costs and increased productivity brings the dramatic benefits of the Internet to work for engineers and scientists. They can use computers with minimal effort to easily share data with their peers across the street or across the world. Engineering students in classroom laboratories can use a computer-based instrument to interact with and learn from equipment in labs and manufacturing facilities that lie miles away from the classroom or laboratory.

Objective Of This Book

While National Instruments products have a wide range of applications, this book provides a broad outline of the industries that have especially embraced computer-based measurement and automation. In the telecommunications, semiconductor, automotive, and biomedical industries, the use of PC-based measurement and automation tools is a growing trend that meets the demands of these industries.

National Instruments and its customers continue to develop application-specific tools for these fast-changing industries. New solutions that closely match industry needs are constantly being developed. This book includes a small sample of some of the latest work done with National Instruments products. The papers published in this book demonstrate how National Instruments software and hardware can solve many different problems. Perhaps these papers will inspire instructors, scientists, students, and hardware and software developers who work in the fields discussed in these pages to find new ways to solve their own unique problems.

Organization Of This Book

This book is divided by industry into five main parts: Automotive, Biomedical, Semiconductor, Telecommunications, and General Test. Each section breaks down the general trends in each industry and describes how National Instruments can meet the unique challenges in each area. The book then presents papers written by users of National Instruments tools.

Automotive Test

Why is Automotive Important?

The automotive industry is one of the largest in the world today. Global vehicle production should rise 1.66 percent in 2000, according to PricewaterhouseCoopers AUTOFACTS, which estimates that approximately 54.06 million cars and trucks will be built worldwide in 2000, up from 53.18 million in 1999. This growing industry demands competitive manufacturing and testing to keep up with current market trends.

What are the Present Trends and Challenges?

Automotive manufacturers and suppliers face increasing consumer demands, stricter government regulations, and heightened competition within the industry. To compete, they must quickly design, manufacture, and test systems and components.

This means automotive test equipment must be faster, more flexible, and more powerful.

To meet these needs, automotive manufacturers and suppliers turn to computer technology to build and test their products. As a result, the following trends have developed in the automotive industry:

- **Growing number of electronic components and subsystems** — Manufacturers install an average of 40 electronic control units (ECUs) in today's automobile engine. ECUs control everything from fuel mixture, to braking systems, to interior climate.

  The paper "Electric Vehicle Inverter Durability Test Stand" details an example of electronic testing. It demonstrates an automated test stand that aids in durability research and development studies on electric inverter units.

- **Shorter and faster product design cycles with immediate turnaround** — Cars now can be designed and created faster than ever before. Customized and just-in-time manufacturing are coming into focus for automobile manufacturers.

  "Automotive Audio Test System" describes a PC-based audio test system that lets manufacturing and repair facilities analyze automotive audio equipment. The system requires high-performance capabilities for end-of-the-line production testing.

- **PCs are the standard for test and control** — The PC is the most flexible and capable platform for automating test. Process improvements for data transfer speeds make it the platform of choice.

  The "In-Vehicle Data Acquisition, Transfer, and Real-Time Processing" comprehensive driver warning application details the use of in-vehicle data acquisition and real-time analysis.

- **Test systems must provide more in-line processing for immediate data reduction and validation** — Immediate test results are possible today. For example, a manufacturer can perform a bumper impact test and then immediately print the results. At one time, this took weeks to complete.

  The application in "Ever Take a Picture of a Pothole From a Moving Truck?" describes an integrated vehicle data acquisition and vision system that measures the loading inputs and the resultant forces on cargo in truck transportation.

- **Reduction in number of warranty returns (higher reliability)** — Automakers strive to reduce warranty returns, and the PC makes better test and verification messages possible.

"PC-Based Control of a Gasoline-Fueled Burner Aging Test Stand to Simulate Engine Exhaust" describes an "oil-less" gasoline exhaust test of automotive exhaust in after-treatment components. The control system monitors two temperature safety levels and shuts down the system if either safety level is exceeded.

What are the Future Trends and Challenges?

Future trends and challenges in the automotive industry include:

- **Driver safety systems will improve** — New systems for driver safety will become standard. This includes air bags, security protectors, collision avoidance controls, and intelligent highways.

- **The use of wireless communications will increase** — The growing field of communications, particularly wireless and satellite, leads to automated toll collection, video and audio traffic sensors, and navigation systems, such as the Global Positioning System (GPS), by which drivers map their location and route with the aid of satellites.

- **The number of electronic control units (ECUs) will grow** — Vehicles will become increasingly sophisticated in electronics, primarily through the ECUs installed.

- **Vehicle bus communications and standards will emerge** — Vehicle bus communications will become more sophisticated causing new standards to emerge.

- **The use of active mechanics will increase** — This includes electronic cars and mechanical and electrical/electronic devices in systems such as antilock brakes, controlled suspensions, load-sensing suspensions, brake-by wire controls, and modular production plants.

How does National Instruments Fit In?

With competition high and technology forcing businesses to become more efficient and produce higher quality products, automobile manufacturers and suppliers find that PC-based measurement and automation systems offer several advantages for test. The trend toward tighter design cycles and the need for immediate results demonstrate that PC-based systems are the ideal choice for lower-cost, high-quality, flexible solutions.

As the quality and power of the PC continues to grow, so do the technologies and solutions National Instruments offers in the automotive industry. Whether designing a dynamometer test stand, wheel bearing test, or an airbag test system, computer-based measurement and automation delivers the power and flexibility to handle any application. With National Instruments hardware and software, engineers and scientists have industry-proven solutions to meet their toughest challenges.

National Instruments delivers one test platform for the entire automotive enterprise. The flexible, high-performance virtual instrumentation platform delivers scalable solutions from the laboratory, to the factory, to the field. Powered by PC technologies, our virtual instrumentation platform reduces cost, increases power, and delivers scalability. National Instruments plug-in and network measurement components meet a wide range of measurement and automation needs, and reliance on the company's industry standards simplifies maintenance and increases the flexibility of your solution. As the power and performance of the PC increases, so does the number of automotive applications for the virtual instrument platform.

Electric Vehicle Inverter Durability Test Stand

Darren Scarfe
Project Scientist
V I Engineering, Inc.

Doug Hornok
Project Engineer
V I Engineering, Inc.

Products Used. LabVIEW™ 5.1, 2- PCI-6602 counter/timers, PCI-6503 digital I/O, SC-2051 breakout board, SC-2062 relay board, CA-1000 enclosure.

The Challenge. Developing an automated test stand to aid in durability research and development studies on electronic inverter units, which are used to convert DC voltage to 3-phase AC voltage for applications within electric vehicles.

The Solution. Using a LabVIEW-based application under the Windows NT operating system to allow for continuous testing and monitoring for 45 days or more. PCI-6602 counter/timer boards were used to monitor pulse train overlap during thermal cycling. In the event of an external power failure during testing, the LabVIEW application is memory capable, and can therefore recall all associated information from the crashed test run. Once the system has regained power, the software will continue testing from the position of power failure and thereby complete the entire test cycle.

Introduction

V I Engineering, Inc. was contracted by Ecostar (a joint venture between Ford, DaimlerChrysler, and Ballard Power Systems) to develop an automated test stand for testing inverter units. "Ecostar's mission is to provide clean and efficient solutions for using energy in personal transportation and related markets. Ecostar Electric Drive Systems is a world leader in the design and development of electric drive systems for automotive and stationary power applications." Due to the long testing times (over 45 days) that are associated with each inverter durability test run, the Windows NT operating system along with additional memory functionality implemented within the LabVIEW application, allows for uninterrupted testing and comprehensive data file management. The test stand is used as a key-life tester to monitor for pulse overlaps that may occur on adjacent lines within a single inverter. The test stand is capable of monitoring three inverters with the ability to expand to five inverters in the future. A custom designed circuit board enclosed in a CA-1000 is used to bring the –5 V to 14 V square wave pulses down to 0-5 V TTL pulses. This permits the use of a PCI-6602 counter timer board so that pulse overlaps can be recorded with a resolution of up to 8 ns. with prescaling.

System Description

The LabVIEW application consists of a complete rack-mounted data-acquisition system and a test control computer. Two National Instruments PCI-6602 counter/timer boards are used to enable the 9 channels required for data acquisition on three inverter units. A PCI-6503 digital I/O board broken out using a SC-2051, and cabled to a SC-2062 relay board, controls power to the inverters. A National Instruments single height rack mount kit is also used to package the breakout boards within the industrial rack. The data acquisition signals from the inverters are run through a custom designed circuit board enclosed in a CA-1000, which converts the signals down to 0-5 V TTL levels, along with the implementation of AND gates to pre-determine whether an overlap has occurred. If an overlap has occurred, then a 5 V signal is routed directly to the PCI-6602 counter/timer board, which is used to keep a running record of the number of overlaps that have occurred. Using the PCI-6602 allows the system to determine signal overlaps with a minimum time-base resolution of 8 ns. with prescalers. Any pulse overlaps are then displayed on the run test screen and also logged to a data file. There are typically approximately 10,950 power on/off cycles; however, the number of power cycles can be user defined, while thermal cycling is done continuously between user-defined temperature values.

When the software application is executed, the user is presented with a login screen. The user is prompted for a password, and is then granted operator, technician, or administrative privileges based on the password entered.

Chapter 1 • Automotive Test

Figure 1-1
The main control screen

This screen also appears when the re-login button is pressed from the main control screen shown in Figure 1-1. The system will allow a maximum of three login attempts before it will close down the program.

The Low Voltage Key Life Tester will initiate testing of a set of three inverters designed to convert DC voltages of ~300 VDC to ~400 VAC, 3 phase power, for powering an electric vehicle motor. There are two styles of inverters, a main inverter board (MIB) and a starter inverter board (SIB.) Each style operates in the same fashion, and has effectively the same inputs and outputs, therefore this test system operates either style of inverter identically for each test.

The inverters are run in a low-voltage mode, which will not successfully operate the inverters in a full power mode, but will create pulses on each of the output lines for each inverter. Each pulse has a width of 8 μs, and no pulse on the supply side of any phase can overlap with a pulse on the return

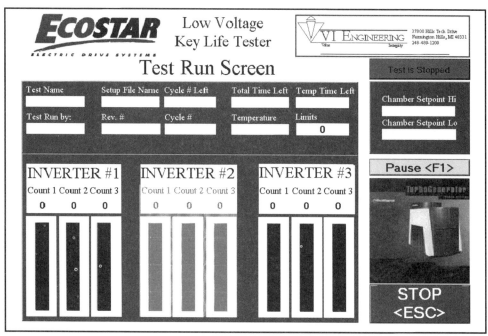

Figure 1–2
The run test screen. All setup information and record counting for cycle numbers are displayed on the run test screen. If an inverter is not selected in the test setup, then it appears grayed-out.

side on any phase. This pulse is created by applying 13.8 VDC via relays using a manual power supply for each inverter.

The relays are triggered using software control commands via a PCI-6503 digital board connected to the SC-2062 relay board. Each relay remains engaged for the period specified as the cycle on time duration in the software. The power is temporarily removed, which is defined as the cycle off time, then re-engaged to begin a new cycle. It is important to maintain contact closure during the entire on cycle to best simulate driving conditions. The inverters are triggered to low-voltage start by a user-specified time (default is five minutes), and maintain that voltage during the entire cycle with a goal to power cycle each inverter a user-defined number of times (default is 10,950). The run test screen is shown in Figure 1–2.

While each unit is being cycled, a temperature chamber is controlled by the RS-232 communication protocol to a 4800 Series Thermotron controller. The chamber begins the first cycle at a user defined high setpoint temperature. The temperature will be maintained for a user-defined time (in hours). Power to the inverters are then cycled on/off, in relation to user defined time intervals for both the on and off time. During the on time, the PCI-6602 board monitors the channels for possible overlaps and records the overlap count in a data file. The data file includes all setup information within a header, along with the date/time, temperature, cycle number, and the channel at which an overlap was recorded for each overlap. The data file is then saved in tab-delimited ASCII format for viewing in Microsoft Excel. At the end of each thermal cycle, the power cycling of inverters is temporarily halted while the chamber is given the new low setpoint. After the low setpoint has been successfully reached, testing resumes and the temperature will remain at this low setpoint for the user-defined time. At the end of the cold cycle, the testing will be halted temporarily, the original high setpoint will be commanded, and after reaching this level, testing will resume. This complete temperature cycle will take place until the specified number of test cycles of the inverters has been reached.

Timing information of the pulse overlaps and associated information are logged as scalar values. Any pulse on a supply side of a phase which overlaps with the pulse on the return side, is flagged as a failure. To save hard disk space, only raw data associated with a failure of a channel will be saved, along with scalar information from all channels for each cycle. If a test is interrupted for any reason, including a power failure, a computer crash, or user intervention, it can be resumed where it left off. The program is capable of automatically keeping track of the cycle number, current temperature, cycles left, and all current overlaps. This is done by using a memory data file system which records whether the current test cycle has been fully completed. If the test cycle has not been run to completion, then when the user initiates the next test run, the computer will default to the previous test run and can then re-call all information required to continue testing from the last position in the cycle time for the previously halted test. If the user wants to run another test, then there are options to either continue the old test, or reload the setup file and re-start a new test.

Results

V I Engineering has implemented a PC-based test stand coded in LabVIEW that aids in the research and development of electronic drive systems inverter units. The software allows for the complete automation of a single testing cycle of 45 days or greater. The operator is merely required to initiate testing and possibly restart the LabVIEW application in the event of an external power failure. The memory function of the software allows for the continued testing starting at the point of power failure in order to complete an all-inclusive test run and data file package.

■ Contact Information

Darren Scarfe

> Project Scientist
> V I Engineering, Inc.
> 37800 Hills Tech Dr.
> Farmington Hills, MI 48331
> Tel: (248) 489-1200
> Fax: (248) 489-1904
> E-mail: dscarfe@vieng.com

Doug Hornok

> Project Engineer
> V I Engineering, Inc.
> 37800 Hills Tech Dr.
> Farmington Hills, MI 48331
> Tel: (248) 489-1200
> Fax: (248) 489-1904
> E-mail: dhornok@vieng.com

Machine Vision in Automotive Instrument Manufacturing

Ganesh Devaraj, Ph.D.
Managing Director
Soliton Automation Private Limited

S.B. Rajnarayanan, M.E.
Senior Project Engineer
Soliton Automation Private Limited

A. Senthilnathan, B.E.
Project Engineer
Soliton Automation Private Limited

Products Used. IMAQ™ 1408, LabVIEW, IMAQ Vision, Fuzzy Logic Toolkit, PC-DIO-24 Low-Cost Digital I/O and PCI-6025E, Low-Cost PCI E Series Multifunction I/O.

The Challenge. Developing a flexible and robust machine vision system for automated calibration of gauges and inspection of instrument clusters in the assembly lines of an automotive instrument manufacturer.

The Solution. Creating powerful and robust image processing software to read gauges developed in LabVIEW using IMAQ Vision. This software, along with DAQ and IMAQ hardware from National Instruments, has been deployed in multiple applications ranging from calibration of individual analog meters (gauges) to inspection and final test of instrument clusters.

Introduction

A leading automotive dashboard instrument manufacturer (Pricol, India) wanted a machine vision system to automate the calibration and inspection of analog (needle-based) speedometers and instrument clusters in their assembly lines. They manufacture complete automotive instrument clusters including a wide variety of speedometers. They wanted a system that would be able to handle all of their present meters and also ones that they would be manufacturing in the future. Basically, this meant that the vision system would have to be capable of reading any kind of analog speedometer without any code changes.

Since the system was to be installed in a factory assembly line, Pricol wanted the system to be tolerant to vibrations. The system would have to automatically correct for small rotations and lateral movements of the meter with respect to the camera. This also meant that the jig holding the meter needed to have more dimensional tolerance without affecting the accuracy of the reading.

The other requirements were:

- Easy configuration of the system for new types of meters
- Automatic recognition of different meters that have been pre-configured, allowing the possibility of different lines, assembling different meters, to use the same calibration station
- Fast and accurate readings
- Integration of other data acquisition and control functions into the system
- Incorporation of Statistical Process Control (SPC) routines and connectivity to a database

Dedicated machine vision systems used in the industry do not meet all these requirements. We determined that a PC-based machine vision system would be able to provide all these features and the required flexibility.

Objective

Recognizing that a machine vision system meeting similar requirements would be useful to both speedometer manufacturers and manufacturers and testers of analog meters, we started with the objective of building a general-purpose system that would be able to handle a wide range of analog meters. As the design started maturing, we realized that we could build an extremely general-purpose system. We also found that LabVIEW and IMAQ Vision provided all the features necessary to implement the design very quickly, including the necessary image processing algorithms. This allowed us to concentrate on developing the needle-detection algorithm, which was the key to the gauge image processing software. The algorithm had to detect the needle in the image of the meter (gauge) and determine the value indicated for any analog needle-based meters, while correcting for rotations and offsets in the acquired image.

System Description

The system we developed, GaugeVIEW machine vision software, formed the core of our application for gauge inspection and calibration. It has been used to develop a complete test and calibration application performing two distinct functions on the production line—calibration of individual meters and the final testing of instrument clusters.

Using the Fuzzy Logic Toolkit in LabVIEW, the GaugeVIEW system was integrated with a fuzzy logic controller for the speedometer calibration application. The calibration utility is designed to calibrate both mechanical analog meters that use magnets and electronic analog meters. In the case of magnetic meters, it controls the demagnetizer, which brings the permanent magnet in the meter to the required level from is initial over-magnetized state. In the case of electronic analog meters, it modifies the calibration parameters in the Electrically Erasable Programmable Read-Only Memory (EEPROM) of the meter.

The GaugeVIEW program is capable of reading multiple meters from a single image of the cluster or read single meters in the cluster with a high degree of accuracy. The cluster inspection utility incorporates these features of GaugeVIEW to check the calibration of all the gauges in the cluster and also check the indicator lamps, the general illumination, and the odometer in the cluster during final inspection.

The primary components of the system are:

- PC (Pentium running Microsoft Windows NT/98/95)
- Cameras
- IMAQ PCI-1408 Image Acquisition Card
- GaugeVIEW gauge image acquisition and processing software
- Calibration Utility
- Cluster Inspection Utility.

The basic operation of the system is as follows:

1. An image of the meter is captured from the camera through the image acquisition card and loaded into the computer's memory.
2. The image is analyzed using software, and the needle position is detected.
3. The meter reading is output in engineering units (MPH, KMPH, RPM, PSI, etc.).
4. The meter reading output from the GaugeVIEW software is fed to the calibration utility or the cluster inspection utility, as required.

The GaugeVIEW meter reading software written in LabVIEW contains two main modules: configuration and read. In order to include a new meter type in the setup, it has to be taken through the configuration process. Once configured, the meter is read without any user intervention using the read module. The main features of the software are:

1. Powerful image processing tools are provided in a user-friendly interface (Figure 1-3) to isolate fiducials in the image and to improve the contrast between the needle and the background. The fiducials are used to correct for rotations and shifts caused by vibrations during run time.
2. Two robust needle-detection algorithms are provided. The user can choose to use either one. To improve accuracy both can be selected, in which case the system averages the results from both algorithms.
3. An extensive calibration utility (Figure 1-4) is provided to map the needle position (degrees) to engineering units. Non-linear graduations on the dial can be easily handled using the polynomial curve-fitting utilities provided in GaugeVIEW.
4. Complete documentation and online help is available at each step of the configuration process, and numerous productivity enhancement features are also provided to reduce the time taken for configuration.
5. A very user-friendly interface is provided to make the configuration process very simple. Figure 1-3 and Figure 1-4 show two of the setup screens used for the configuration of a new meter.

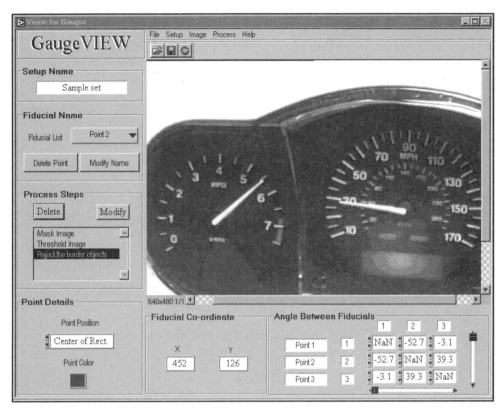

Figure 1-3
Fiducial setup screen

GaugeVIEW is optimized such that the configuration module is very user-friendly and the read module is very lean and fast. On a Pentium II 266 MHz processor with 64MB of RAM, an image can be processed in less than 50 msec with an accuracy of better than one degree on a typical meter. The algorithm can correct for on average of ±5 degrees of rotation and 4 mm (20 pixels) offset. These values depend on the meter dial layout and size, but the values given are typical.

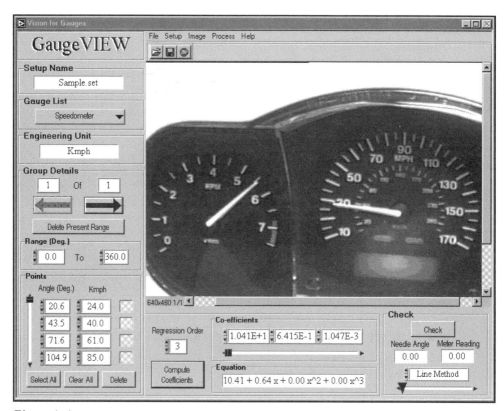

Figure 1-4
Calibration setup screen

Calibration Utility

Calibration of mechanical analog speedometers involves the control of a demagnetizer and a DC servo motor. Mechanical analog speedometers use a permanent magnet to drive the needle. A demagnetizing coil (electromagnet) is positioned close to the permanent magnet of the meter when the meter is mounted on the calibration stand. During calibration, a controlled AC current is passed through the demagnetizing coil to slightly demagnetize the permanent magnet. The magnet is manufactured in an over-magnetized state so that it can be demagnetized to the right level during calibration. The demagnetizing pulses are continued until the calibration is complete.

For example, before calibration, the speedometer will indicate a higher speed than the actual value corresponding to a given RPM due to over-magnetization. The meter could indicate 80 mph instead of the correct value of 50 mph. Successive application of demagnetization pulses would then reduce the reading, say from 70 mph to 60 mph to the correct value of 50 mph.

Because the demagnetization response of the different speedometers was not simple to model, we implemented the control algorithm in fuzzy logic to most closely mimic the actions of a skilled human operator. Using the Fuzzy Logic Toolkit from National Instruments, the implementation was done very quickly. At each step, the AC voltage was applied to the demagnetization coil and the duration of the application was determined based on the difference between the present speed and the target speed. A PC-DIO-24 low-cost digital I/O board from National Instruments was used to interface with the external circuitry to select the AC voltage and also to apply the demagnetization pulse of the right duration.

The utility has a graphical user interface (Figure 1-5), which shows the current status of the test on the computer screen. The software permits the user to set the operator information and select the meter type, description, etc. The screen also shows indicators for the calibration completion and results. During the calibration, the screen will show the current action being performed although there is no operator present at the calibration station. The operator involved in the previous step in the assembly process mounts the meter and presses a start button. The operator who handles the next step in the assembly process picks up the meter after calibration. The user interface provides information which is useful when the process is being monitored by the operator.

A performance comparison of the earlier manual system and the new automated system is given in Table 1-1. There is substantial improvement in accuracy, speed, and the elimination of errors due to human neglect or fatigue.

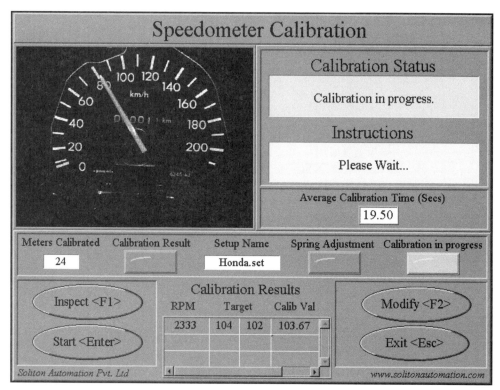

Figure 1-5
Calibration Utility graphical user interface

Table 1-1   Performance comparison of Manual and Automated Calibration

Performance Parameter	Old (Manual) System	New (Automated) System
Accuracy	± 2.5 KMPH	± 1 KMPH
Speed	16 to 40 seconds (depends upon fatigue)	18 to 22 seconds
Calibration Error due to Over-Demagnetization	3 percent of the meters calibrated	Less than 1 percent of the meters calibrated

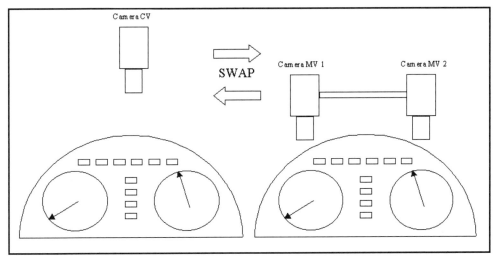

Figure 1-6
Schematic of the cluster inspection station

Cluster Inspection Utility

The cluster inspection utility is designed to perform inspection and final testing of complete automotive dashboard instrument clusters. A typical cluster includes speedometers, odometers, tachometers, fuel and temperature gauges, and high beam/low beam and turn signal indicator lights. The cluster inspection module consists of three cameras—two meter view (MV) monochrome cameras for accurate inspection of the speedometer and tachometer and one cluster view (CV) color camera to inspect the entire cluster. There is a mechanism to move the cameras into place over the meters. The cluster inspection module is designed to inspect two clusters simultaneously. Initially the cluster view camera captures the image of the cluster on the left, while the two meter view cameras capture close-up images of the speedometer and tachometer in the second cluster on the right (Figure 1-6). After the tests have been completed, the cameras are interchanged and the process is repeated.

Camera CV will view the entire cluster for the inspection of lamps in the cluster, intensity and color of the lamps, and needle movement of fuel and temperature gauges. We are using the still color capability of the PCI-1408 image acquisition card to obtain the color image of the cluster. Cameras MV 1 and MV 2 are used to inspect the calibration of the speedometer and tachometer in the cluster. Simultaneously, the calibration of the tachometer and the speedometer are checked. The output of MV 2 is also used to check the odometer.

One of the main advantages of using a LabVIEW-based virtual instrumentation system is evident here. Since LabVIEW integrates data acquisition and image processing capabilities, we were able to provide the control signals to the cluster (DAQ) and check the response (image processing) without having to work with multiple development environments. The outputs from the MIO board were used to switch on the voltage to the different indicators and to provide the stimulus to the gauges (analog signal to the DC motor driving the speedometer, and digital pulses to the tachometer). Simultaneously, the image of the cluster and gauges were acquired through the IMAQ board to check the response.

The cluster inspection utility is designed to carry out the following tests on the cluster:

- Checking the position of the lamps in the cluster (for example the bulbs should not be interchanged)
- Checking the intensity of lamps present in the cluster
- Testing the calibration of the speedometer and tachometer
- Testing for the speedometer and tachometer pointer wavering
- Damping of the speedometer needle – time taken for the needle to return from 200 KMPH to 0 KMPH when the excitation is suddenly removed
- Testing the odometer alignment and rolling smoothness using optical character recognition (OCR) and other image processing algorithms
- Testing the needle movement of fuel and temperature gauge – this is just to verify that the meters are operational.

Results

The machine vision system for gauge inspection and calibration is a full-featured PC-based system used to read a wide range of analog meters. It is also a versatile system that is ideal for various applications in the automotive instrument manufacturing industry. Since the software is written using LabVIEW, it can seamlessly be integrated into test applications along with a host of other National Instruments hardware and software tools. We were able to build a very powerful and user-friendly system in an extremely short time due to the tremendous productivity gains we achieved by using National Instruments software and hardware products.

■ Contact Information

Ganesh Devaraj
Managing Director
Soliton Automation Private Limited
Classic Towers, 1547 Trichy Road
Coimbatore 641018, INDIA
Tel: +91 (422) 302374
Fax: +91 (422) 302375
E-fax: (305) 723-0908
E-mail: ganesh@solitonautomation.com
http://www.solitonautomation.com

Tire Rolling Resistance Measurements Using LabVIEW and VXI

Richard DeBin
Project Engineer
Datappli, Inc.

Products Used. LabVIEW, MXI cards, PID Control Toolset, VXI chassis.

The Challenge. Developing a comprehensive software application to automate an existing tire dynamometer. The software application must reduce technician involvement, minimize cycle time, and increase repeatability.

The Solution. Using products from National Instruments, including LabVIEW, PID Control Toolset, VXI chassis and MXI cards ensured that the best quality components were used to develop a software application that could handle complex control issues while maintaining an intuitive operator interface to control the whole test.

Introduction

The original dynamometer and test procedure was very labor intensive. A technician was first required to mount the tire on to a positioning carriage and fill the tire with air to the appropriate pressure. Next the technician lowered the tire into place by moving the positioning carriage down. Then the technician had to start up the drive motor and adjust its speed. At this point the technician could apply a load to the tire by moving the positioning carriage up or down and by adjusting the air pressure in the loading bladders. A proprietary computer system would then continuously measure the rolling resistance of the tire and determine when a state of equilibrium had been achieved before taking the final measurements. The technician then had to set up the next loading condition in the test sequence by repositioning the carriage and adjusting the air pressure in the loading bladders.

Datappli developed a LabVIEW-based application which utilizes a C-size VXI chassis and a VXI-MXI/AT-MXI interface to a personal computer to fully automate the testing procedure listed above. This application also provides calibration routines that accommodate the company's ISO 9000 requirements, as well as ways to graphically view one or multiple test results. With this new program in place, the technician needs only to mount the tire on to the positioning carriage. The program automates everything, including filling the tire with air to the proper test pressure. This high level of automation significantly reduced the involvement of the technician, which increased the test repeatability and decreased the test completion time.

Software

Datappli developed the data acquisition and control software using the National Instruments LabVIEW programming environment. LabVIEW's graphical programming environment was very useful in reducing development time and thereby development costs. This graphical programming environment makes it easy for novice programmers or engineers to learn quickly how to program the complex control algorithms and analysis routines necessary for the test routines. Toolkits available for LabVIEW provide pre-written sections of code that are useful in programming such tasks as regulation of tire pressure and loading.

$$Output = (\pm)K_p * e + K_i \int e \, dt - K_d \frac{dPV}{dt}$$

The regulation of the tire tread loading was complicated by the use of air bladders and the tire positioning carriage. The air bladders introduced a significant control lag into the system. Furthermore, the positioning carriage had to be placed at a certain height according to the requested tire tread loading. The height of the carriage system sometimes required the PID control to reverse the action of the control calculations shown below. When the PID control action needed to be reversed, the proportional gain needed to be inverted without causing a huge and sudden change in output.

The PID Control Toolset for LabVIEW added the flexibility needed to address all of these issues. It saved development time because the prewritten code could be easily inserted into the data acquisition and control code that was being developed. The system can regulate tire tread loading to within

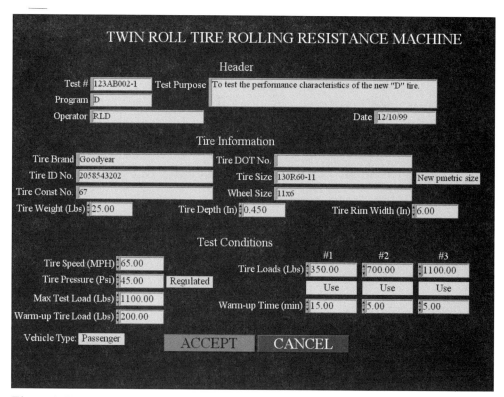

Figure 1-7
Setup screen used to define test parameters

plus or minus two pounds. The original software did not provide any regulation. The technician had to set the loading manually and adjust it throughout the test. The tire pressure can also be regulated to within plus or minus one pound per square inch. As with the tire tread loading, the original software did not provide any regulation, and again the technician had to manually adjust the pressure.

This application allows the technician to define test setups quickly and easily using only one setup screen, as seen in Figure 1-7. The technician can execute up to three test conditions with little or no intervention. In addition to automated tests, the technician can remotely operate the system from the computer. The data from the automated tests is stored on the hard drive and can be reviewed on the computer monitor, as well as in printed reports generated by the system at the technician's request. To ensure accuracy and repeatability, the sensors can be calibrated and the results recorded with

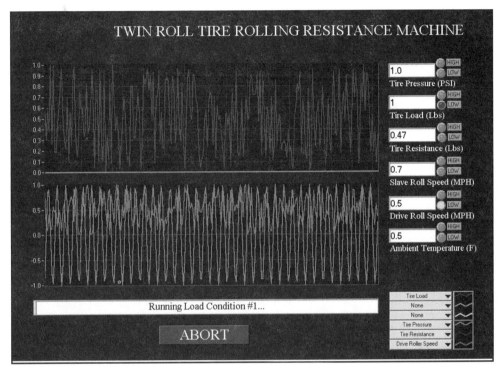

Figure 1-8
Example of graphical data display

every test data file. The machine losses, the added loading and drag created by friction within the dynamometer, for the complete system, can be measured by the LabVIEW based data acquisition system. The system then corrects the measured data for these losses.

Another benefit of using LabVIEW is that the graphical user interface (GUI) provides an easy-to-understand way of showing generous amounts of data as seen in Figure 1-8. LabVIEW's many different graphical indicators allow data to be represented in a more familiar way, such as dials, meters or chart displays. These indicators copy the look of bench top test equipment, such as a multimeter. Control options are also available through the graphical user interface in an easy-to-understand format. This is due to the graphical indicators and controls representing the data not as mere numbers but more pictorially.

The LabVIEW software allowed us to keep the main interface simple and intuitive. By applying a "tree" structure to the screen layout the manual

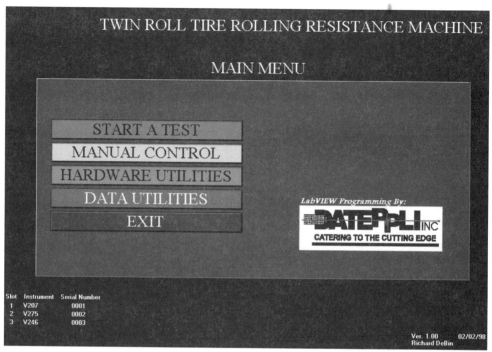

Figure 1-9
Main menu layout

operation of the system and automatic testing operations were kept within the first level (Figure 1-9). This kept the screen navigation to a minimum, which is important when designing a program interface that is simple and intuitive. The importance of a simple and intuitive interface is realized most often when training new technicians. The simpler the program is to understand, the quicker someone can be taught how to use it.

Hardware

One of the most challenging of the system's hardware capabilities to provide was the range of actual tire tread loading that could be applied. The tire tread loading is provided by a system that includes two air-inflated bladders and a complex tire positioning carriage system. The tire tread loading conditions can be set to a static load ranging from 50 to 2,500 pounds and held to within two pounds over the length of each of the three test conditions. The system can also reach tire speeds of up to 80 miles per hour and tire pressures up to 100 lbs per square inch. The tire pressure can also be regulated to a constant value or be sealed off to simulate the actual use as seen on a vehicle.

The hardware consisted of a small control cabinet to control the dynamometer hardware and a Pentium-based computer. This replaced the customer's sizable, proprietary computer system. The decrease in hardware size offered an opportunity to install control hardware for other dynamometers in the limited space that is available. A National Instruments thirteen slot C-size VXI chassis situated inside the control cabinet provided sensor conditioning and control signals to the tire dynamometer. Based on the customer's specifications for the data acquisition hardware, the following modules were chosen for this system. The first slot has a VXI-MXI module used for controlling the VXI chassis. The computer and VXI chassis communicate through a National Instruments AT-MXI card installed in the computer and the National Instruments VXI-MXI mainframe extender. The second slot has a voltage input module, which converted signals from the next slot to digital data. The third slot contained the signal conditioning module that accepted the signals from various sensors. Slots four and five are empty. Slot six is a voltage output module used to generate voltage signals for controlling the tire speed; tire pressure and tire tread loading of the twin tire rolling resistance dynamometer. Slot seven is a multimeter, which converted signals from the next slot. Slot eight is a thermocouple relay module for reading a RTD (resistive thermal device) which measured the dynamometer room temperature for monitoring purposes. Slots nine, ten, eleven, and twelve are empty. Slot thirteen is a digital I/O module for both reading and controlling the status of the twin tire rolling resistance dynamometer hardware. Among the items controlled by the aforementioned digital outputs are emergency stop control circuits, drive motor engagement and tire air handling valves. The empty slots are available for future expansion of the system.

Results

National Instruments software and hardware provided the tools necessary to make the system upgrade simple and quick. The system itself is efficient, easy-to-use, and repeatable. The technicians only have to mount the tire into place, enter the test setup information and click on start. They no longer have to start up the drive motor and apply a load to the tire—the LabVIEW based data acquisition system does both of these now. This makes the newly upgraded system highly automated, thereby minimizing test cycle time and allowing the operator to oversee the progress of two or three other machines simultaneously. This high level of automation also decreases human involvement and thereby increases repeatability of the tests.

■ Contact Information

Richard DeBin
Project Engineer
Datappli, Inc.
21314 Melrose Ave.
Southfield, MI 48075
Tel: (248) 353-5212
Fax: (248) 353-4913
E-mail: rdebin@datappli.com

PC-Based Control of a Gasoline-Fueled Burner Aging Test Stand to Simulate Engine Exhaust

Cynthia C. Webb
Senior Research Engineer
Southwest Research Institute

Earl F. Quillian
Research Assistant
Southwest Research Institute

Products Used. LabVIEW, SCXI™, and E Series DAQ.

The Challenge. Developing a test stand that provides oil-less gasoline exhaust at a precise air-to-fuel (AF) ratio for aging and testing automotive exhaust after-treatment components. Test stand control system requirements included: operator-controlled component actuation, closed-loop AF ratio control, and acquisition, analysis, and storage of data over extended periods of time.

The Solution. Creating stoichiometric (chemically correct balance of fuel and air) gasoline-fueled burner exhaust test stand with a supplementary oil injection system and supplemental control system to run aging cycles and perform closed-loop burner AF ratio control. Control was achieved via an E Series DAQ board in a PC running a LabVIEW-based user interactive program, and an external SCXI chassis with SCXI 1120s for data acquisition.

Introduction

The Environmental Protection Agency (EPA) now requires that automotive emission control components function for 100,000 miles. This requirement, coupled with the lower emission standards, places severe demands on catalytic converters and other exhaust after-treatment devices. Fuel and lubricating oil may contain small amounts of sulfur, phosphorus, MMT, calcium, or zinc. These constituents can degrade catalyst performance to the point where the vehicle no longer meets its designated emission standard.

A screening procedure was needed to evaluate the effect of fuels, oils, and candidate additives on catalyst performance. If the evaluation procedure were to include the use of a gasoline-fueled engine, it would be difficult to separate effects due to fuel additives from additives in the lubricating oil because both can be present in engine exhaust. Also, as engines wear, the effect of combustion chamber deposits and material wear can alter the amount and composition of each additive in the exhaust. Therefore, because the use of an engine may confound the impact of additives on catalyst performance, Southwest Research Institute (SwRI) developed the Fuel/Oil Catalysts Aging System (FOCAS) that would allow additives to be evaluated without the use of an engine.

System Design

Each of the test stand subsystems were designed and developed by SwRI. The test stand subsystems consist of the following:

- A gasoline-fueled, stoichiometric (chemically correct balance of fuel and air) burner
- An oil injection subsystem to allow the introduction of oil in a controlled quantity at any location within the exhaust
- A computerized control system to operate the entire test stand.

The steady-state, stoichiometric burner apparatus can simulate the flow of exhaust from a four cylinder engine under a variety of load conditions. The oil injection system provides control over the amount and oxidation state (unburned, partially burned, or completely burned) of consumed lubricating oil. The computerized control system allows the burner system to be operated safely away from the bench. Full-sized automotive catalysts can be tested as the test procedure simulates actual exhaust flow rates for a typical four-cylinder engine. Figure 1–10 shows the burner portion of the system.

The control system consists of a LabVIEW-programmed PC equipped with a touch screen monitor, a multi-function data acquisition (DAQ) card. The PC connects to an SCXI chassis holding two SCXI 1120 multiplexing mod-

Figure 1-10
The FOCAS system

ules, one feed through panel, and an SCXI 1160 relay module to monitor and record system information and control system electronics.

Using the computer interface, the operator can switch power to the blowers and fuel pump and control the air-assisted fuel injectors, burner spark, oil injection, and exhaust bypass with the touch of a finger. System temperatures, a mass air-flow sensor for burner air, and the burner AF ratio are measured and converted to engineering units. The software uses measured data to calculate total exhaust flow and burner AF ratio, and to check conditions indicative of a system malfunction. The burner AF ratio can be controlled as either open- or closed-loop, maintaining the specified AF ratio.

Adjustments can be made to vary the rate of fuel to the burner. Fuel injection rate control is achieved with a "square" waveform generated by a DAQ counter timer (fixed frequency) with a variable duty cycle (pulse) to the fuel injector. Whenever necessary, open loop control is achieved by allowing an operator to enter two setpoint duty cycles. The program can then be switched between the two setpoints every 10 seconds. Closed loop control is achieved by measuring the actual burner AF ratio, comparing the measured value to the AF ratio setpoint, and then adjusting the fuel injector duty cycle to correct for the measured error.

The front panel of the program is designed to allow users to input an aging test and to run the test using a single screen. The test diverts the flow of exhaust through the bypass leg of the test system. Then, each of the test bench components is activated and adjusted.

After the system is adjusted, an operator enters the two safety setpoints and the setpoint AF ratios, diverts the exhaust into the test piece, and begins data acquisition. Data acquisition can be paused and resumed at any point during a test. While a test is in progress (after data acquisition has begun), the program checks the two safety points once per second. Data is held in a buffer gathered at 1 Hz, for 60 seconds and a one-minute average is stored, along with the minimum and maximum values for each channel.

In addition, the average, minimum, and maximum temperatures of the catalyst bed and measured AF ratio are plotted to the front panel, allowing an operator to review the overall stability of the system. Figure 1–11 shows the front panel of the control software. The front panel depicts the layout of the actual test system and the location and value of the measured data at each point in the system.

Due to the inherent danger of leaving a gasoline-fueled burner operating unattended for periods of time, the system uses two built-in safety limits that check exhaust temperature. First, the exhaust temperature must remain greater than the minimum safety setpoint level, or which indicates that the burner is still lit. The second setpoint checks the catalyst bed temperature to verify that the catalyst is not at a temperature that could be detrimental to the experimental part. If either safety setpoint is exceeded, the computer is programmed to turn off all test systems except the burner air and to display a bright red screen describing the condition at which the system was shut down, along with the date and time.

Results

LabVIEW and SCXI were assembled and programmed to create a control system providing a safe and easy method for using a complex and potentially dangerous gasoline burner test bench. Benefits that this controlled aging test bench system can provide include the definitive isolation of additive effects, states of oxidation in lubricating oil, and various fuel and oil effects on catalyst performance.

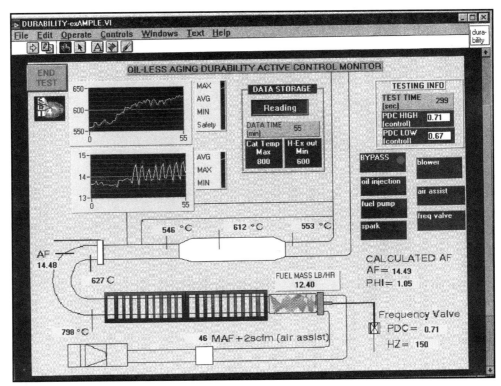

Figure 1-11
Front panel of LabVIEW-based control software for the FOCAS burner system

In addition, the LabVIEW software is flexible, allowing this program to be modified to accommodate future changes in testing requirements. Both SCXI and the DAQ boards possess significant expansion potential including unused analog inputs/outputs, counter I/Os, and digital outputs which allow for easy expansion of the current system's capabilities.

This system has been used to successfully carry out many catalyst aging and poisoning evaluations and has run continuously for up to 200 hours at a time. The safety systems have worked well to protect experimental catalysts systems and to shut down the fuel injection and spark systems in the event of an unexpected burner failure.

■ Contact Information

Cynthia Webb
Senior Research Engineer
Southwest Research Institute
P.O. Drawer 28510
San Antonio, TX 78228
Tel: (210) 522-5873
Fax: (210) 522-3950
E-mail: cwebb@swri.edu

In-Vehicle Data Acquisition, Transfer, and Real-Time Processing

David Hoadley, Ph.D.
Senior Project Scientist
V I Engineering, Inc.

Jeff Siegel, Ph.D.
Senior Project Scientist
V I Engineering, Inc.

Darren Scarfe
Project Scientist
V I Engineering, Inc.

Products Used. National Instruments Products: LabVIEW, SCXI hardware: 1200 Multifunction I/O Module, 1120 Differential Amplifier, 1124 Analog Output Module, 1141 Elliptic 8-Pole Filter Module, 1126 Frequency to Voltage Converter, 1163R Solid-State Relay Output.

Third-Party Products: Panasonic S-VHS VCR, Elmo CCD cameras, Horita MTG-50, Time Code Generator, Trimble Accutime GPS receiver, Aerotron-Repko FSK RF modems, and Black Box Wireless LAN system.

The Challenge. Creating a comprehensive driver warning system using in-vehicle data acquisition and real-time analysis.

The Solution. Using LabVIEW and SCXI for data acquisition and control, communicating between moving vehicles via RF modems and a wireless local area network, and supporting a GPS receiver/time code generator to produce an in-vehicle data acquisition and analysis system.

Introduction

In 1996, 1.8 million rear-end automotive crashes occurred in the United States with approximately 2000 associated fatalities and 800,000 injuries. Forward collision warning (FCW) systems are now becoming available, designed to assist drivers in avoiding such collisions by providing timely driver alerts.

The Crash Avoidance Metrics Partnership (CAMP) is a partnership established by Ford Motor Company and General Motors to undertake joint pre-competitive work in advanced collision avoidance systems. The objective of this project was to define specific crash types that an FCW should address, define a minimum functional specification for such an FCW, and develop object test procedures to evaluate the performance of an FCW system. CAMP's findings were published in 1999 [Kiefer et al.].

In order to proceed, CAMP required a flexible in-vehicle data acquisition, analysis, and control system. To implement the Driver-Warning System (DWS), we used SCXI components to perform the data acquisition and control functions and LabVIEW to provide a graphical user interface. Hardware setup, channel units and scaling, and channel selection and identification are achieved via intuitive front panels. The resulting application provides network communication between two test vehicles; serial communications with a range sensor, VCR, and global positioning system (GPS) receiver; and communication with data acquisition hardware for measurement of onboard sensors.

In order to provide driver warnings, the application required real-time transfer of data between moving vehicles. Local and network retrieved data are used to evaluate user-specified warning conditions that can provide visual, auditory, and/or haptic feedback to the driver.

Data Acquisition

For convenience and portability reasons, the DWS was implemented on a laptop computer platform. We selected SCXI components as an in-vehicle data acquisition solution. The SCXI line of products provided an off-the-shelf set of input modules that met our requirements for analog input and output, frequency to voltage conversion, signal filtering, and digital input and output. The 12-slot chassis contains an SCXI-1200 (data acquisition and chassis controller), SCXI-1126 (frequency to voltage converter), SCXI-1120 (differential amplifier), SCXI-1141 (8-pole elliptic filter), SCXI-1124 (analog voltage output), and SCXI-1163R (solid-state relay) modules. Custom hardware used to power sensors and synchronize the multi-vehicle data acquisition is packaged on an SCXI-1181 breadboard module. Vehicle DC power provides AC power to the system via an inverter.

All of the steps necessary to set up the data acquisition hardware can be taken from within the DWS application software. We chose to use the DAQ Channel Wizard as the basis for data acquisition channel setup. It provides the user with the capability to assign meaningful names to channels, enter voltage to engineering unit conversion information, and provide each channel with a description. The analog channel setup screen is displayed in Figure 1-12. To set up a test, the user selects from the list of available channels those that will be used with the on/off selector in the left column. The function selection specifies which channels correspond to the vehicle acceleration, speed, and separation. These parameters are the ones that are useful for real-time analysis. Channels are displayed by Channel Wizard name, with units and a description. The user can run the NI-DAQ™ hardware and channel configuration utilities to make desired system setup changes without leaving the application. Serial device channels and the data acquisition rate are specified here as well. Some examples of data that can be collected are accelerometer signals, vehicle speed, brake switch, steering wheel position, brake pedal load sensors and distance to the front vehicle.

Test Execution and Data Acquisition Synchronization

Data acquisition for all systems in use begins with the selection of a single control in the master vehicle. In order to make meaningful comparisons between local and remote data, great care was taken to synchronize the acquisition on multiple systems. Before the test begins, data acquisition (DAQ) systems in all vehicles are set to the same nominal acquisition rate and configured to begin acquisition upon receipt of a digital hardware trigger. Control messages are sent over the Black Box 1Mbit/s wireless LAN using TCP/IP protocol to initiate the test on all remote systems. The wireless LAN has an effective range of ½ mile when broadcast using relatively small 18″ antennas. The master system waits for each remote system to report back the status of various hardware components.

Once pre-test steps have successfully finished on all systems in use, the DWS application enables the digital trigger. Data acquisition begins on all systems, triggered by a single external hardware trigger. In the master vehicle, the trigger comes from a Trimble Accutime GPS receiver which outputs a digital pulse synchronized with the start of each second. This pulse per sec-

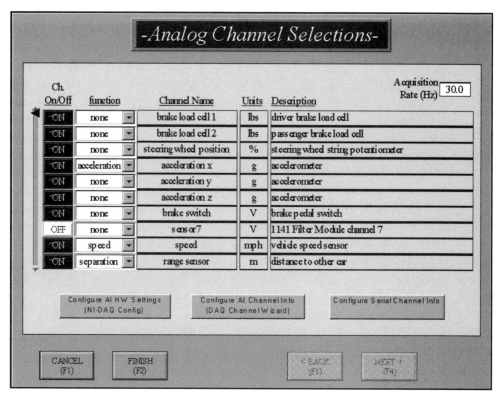

Figure 1-12
Data acquisition channel setup and selection screen

ond (PPS) signal is conditioned by a custom pulse widener and then broadcast to the slave systems by means of Aerotron-Repko RF FSK modems. These devices transmit TTL-level signals at a rate of 2400 baud reliably over distances of up to ½ mile. In this way, we synchronize the start of data acquisition for all systems to within 0.5 mS.

The master system reads the data acquisition start time from a Horita MTG-50 GPS time code generator through a serial port. The time code generator also provides a longitudinal (video) time code (LTC) that is converted to vertical-interval time code (VITC) with a Horita VG-50 LTC-to-VITC converter. The VITC and the video signals from the CCD cameras are recorded onto a video tape using a Panasonic S-VHS VCR. Analog data is acquired at 30 Hz, synchronous with the nominal NTSC frame rate. Typical tests are at most five minutes long; thus, accumulated timing errors associated with the difference between the hardware scan clock and the VCR frame rate are

Chapter 1 • Automotive Test

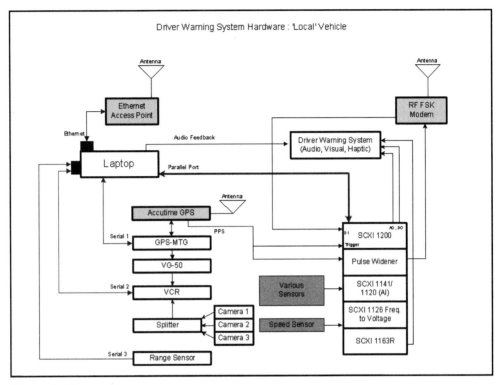

Figure 1-13
Hardware schematic used in the master system and its connections

insignificant—typically less than one frame. The data acquisition in all vehicles is effectively synchronized with the GPS-derived video time stamp, within the limitations of the time-code generator. Test termination is controlled from the master system by TCP/IP-communicated commands. Finally, after data acquisition has ended, data files from all systems are combined into a complete data set. Figure 1-13 is a schematic illustrating the hardware used in the master system and its connections.

Real-Time Analysis

Speed and acceleration data from the remote vehicles are transferred in real time to the master system over the wireless LAN network link. For real-time analysis, we require only the newest data from any remote systems. We therefore use the User Datagram Protocol (UDP) to transfer the data. Unlike TCP/IP, UDP provides no packet tracking features, and therefore data is never re-sent in the event of a temporary loss of network connection. The master system, when and if it receives data from the other vehicles, performs user-defined analysis on data items with matching scan numbers from the local and remote systems. Since the start of data acquisition on all systems is virtually simultaneous, we can provide real-time analysis and thereby immediate feedback of the results of this analysis by user-predetermined visual, audio, or haptic means.

Custom display devices developed by hardware contractors to CAMP, including a High Head-Down Display (mounted on the vehicle dashboard) and a Heads-Up Display (projected into the driver's line of sight), provide visual warnings. These displays are controlled with digital outputs and/or the SCXI-1163R relay output module. Audio warnings are generated by playing user-selected sound files with the laptop sound hardware, amplified through the vehicle's radio. Haptic warnings are generated using an analog output to control a custom vehicle brake booster, generating a brake pulse.

Figure 1–14 shows the analysis setup screen. The user may select up to four analysis formulas, each with a separate warning criterion. The available variables to use for real-time analysis include local and remote vehicle acceleration and speed and the observed distance between the vehicles. The application uses the user-entered formulas to calculate a minimum desired separation between the vehicles. When the actual separation falls below the result of a particular formula, its selected warnings are displayed. The formulas are evaluated with a formula parser, which is a stand-alone product of V I Engineering. Each formula can be set to provide any combination of alert methods, including visual, auditory, and haptic means. The evaluated channels of data are saved along with the local data and can be reviewed with the post-test analysis component of the DWS software. CAMP used the real-time analysis feature to develop and test their driver warning algorithms.

Review Test Data Utility

The interface for review of test data provides the user with a means to view the results of tests. Drawing heavily on the capabilities of LabVIEW, it can graphically display local, remote, or analysis channels. The utility's most interesting feature is that it controls the VCR and reads video time code with a time-code decoder. Because the video recorded during the test and the data

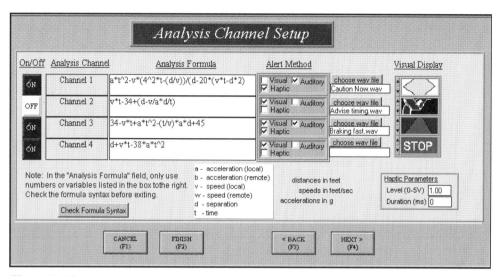

Figure 1–14
Real-time analysis and driver warning setup screen (equations are for illustration purposes only)

were time-stamped, we can determine which frame of the VCR tape corresponds to the current position of a cursor on the graphical data display. When the user chooses a cursor position, the tape can be moved to the corresponding frame, and the data and video can be reviewed in a synchronous manner. Similarly, the graph cursor can follow the current video frame during playback.

Results

V I Engineering has produced an in-vehicle data acquisition, control, and analysis system, which includes real-time data and control signal transfer between moving test vehicles. The ability to create intuitive user interfaces, advanced network communications capability, and ease of data acquisition and control made LabVIEW the development environment of choice for this project. Integration of the DWS posed some unique difficulties, all of which were overcome with the correct mix of V I Engineering expertise with software tools, hardware, and support efforts from National Instruments.

References

[1] Kiefer, R., D. LeBlanc, M.Palmer, J. Salinger, R. Deering, and M. Shulman, *Development and Validation of Functional Definitions and Evaluation Procedures for Collision Warning/Avoidance Systems*, Report Number DOT HS 808 964, sponsored by the National Highway Traffic Safety Administration.

■ Contact Information

David Hoadley, Ph.D.
Senior Project Scientist
V I Engineering, Inc.
37800 Hills Tech Dr.
Farmington Hills, MI 48331
Tel: (248) 489-1200
Fax: (248) 489-1904
E-mail: dhoadley@viengineering.com

Automotive Audio Test System

Grant L. Mobley
Technical Services Manager
Downtown Radio of Denver

Mark Schmitt
Project Engineer
Datappli, Inc.

Products Used. E-series MIO card, GPIB card, LabVIEW, SQL Toolkit, LabVIEW Application Builder, and Component Audio Tester (CAT.2).

The Challenge. Developing a PC-based audio test system for manufacturing and repair facilities to analyze automotive audio equipment. This test system must be cost effective, but still deliver high performance parametric testing capabilities comparable to the costly test systems in use today for end-of-the-line production testing. The system must be capable of performing tests such as wow and flutter, signal to noise ratio (SNR), and sensitivity. The system must also use a standard structured query language (SQL) database to record all test results so that they can be later analyzed.

The Solution. Using National Instruments LabVIEW software, data acquisition (DAQ) cards, GPIB, and SQL Toolkit to collect and analyze audio signals while controlling a GPIB Radio Frequency (RF) generator and a custom RS-232 controlled Component Audio Tester.

Introduction

Downtown Radio of Denver, Inc. (DTR) wanted to develop a PC-based audio test system that could be sold as a stand-alone product. This type of test system has never been available without using many individual pieces of test equipment and customized programming. The basic requirements were as follows:

- The system must be capable of maintaining multiple manufacturer and model combinations in a secure database
- The system must be able to perform the same measurements as more complex analog test solutions
- The system must be affordable at the service level, but perform to the standards of the manufacturing level
- The system must be built in such a manner as to prevent obsolescence and provide not only flexibility, but also expandability to allow for future additions.

DTR needed a Microsoft Windows-based analysis and control program capable of running user definable test suites and reporting the results to different databases. National Instruments LabVIEW software was chosen for its ability to perform complex analysis on the acquired waveforms and its ability to control custom serial devices, GPIB instruments, barcode printers and scanners, as well as National Instruments DAQ boards. LabVIEW's ability to command an SQL database, perform file operations, and provide a friendly, easy-to-use graphical user interface for programming and operation was another key factor in the decision to use it for this application. Datappli Inc. of Midland, Michigan, was contracted to write the programs because of their extensive experience with LabVIEW programming and developing products from customized test systems.

Hardware Design Considerations

One of the primary considerations was to build a system that would allow multiple manufacturers' audio systems to be tested on a single station. This required that the product be built with a universal interface from both a software and hardware perspective. DTR built the CAT.2 vehicle simulation and interface unit to allow any audio product to be connected to the test system. The hardware has many features that are not typically found on a production line tester, such as expansion ports to allow for future enhancements and modifications. It was built from the beginning to be a simulator of the vehicle environment. This meant that many things had to be taken into consideration, such as:

- Supporting Pulse Width Modulated (PWM) and analog dimming
- Monitoring all audio channels
- Supporting multiple, independent, variable DC power supplies
- Testing both passive and active audio products
- Supporting high and low level audio outputs
- Supporting a data interface for the product under test that can handle multiple automotive data buses (J1850, CAN, CCD, and others).

Hardware Design

The customized integrated test system is now being manufactured for and sold by DTR as the CAT.2. The CAT.2 is a full-functioning vehicle simulator. Many additional enhancements were built into the product to allow for things such as an audio monitoring function. This is a function that allows the product under test to run under full load while only allowing a small audio sample to be sent to the speakers. This permits a bench technician to test products that fail under load in a practical environment. The ability to test radio frequency (RF) modulated components is also built into the unit, eliminating the need for many pieces of external test equipment. Other features such as a replaceable protective mylar overlay on the touch screen will help the product to last for many years.

The CAT.2 is capable of providing all the signals necessary to test audio products as if they were actually in a vehicle. This includes high and low level audio, internal amplification, and a data interface for automotive data buses such as J1850, controller area network (CAN), charge-coupled device (CCD), and others. The CAT.2 contains several power supplies for variable B+, ignition, parking lamps, and both analog and pulse-width modulation (PWM) dimming control. All of these power supplies can be set independently of one another. This eliminates the need for costly GPIB controlled power supplies. It also contains several relays for switching in precision internal loads or external speakers and a universal connection port for the product under test. Add this to the ease of programming in the LabVIEW environment and obsolescence is almost impossible.

We planned to use National Instruments data acquisition hardware from the beginning of the project. The back of the CAT.2 has a 68-pin connector that is compatible with the E-series PCI-MIO product line. The LabVIEW RS-232 communication VIs, which are built into the software package, give the program complete control over the CAT.2. A Pentium PC is the center of the system from which the program, written in LabVIEW, controls all the hardware and collects the data. The system also incorporates a laser printer, a barcode printer, barcode scanner, and a GPIB controllable RF generator. The entire product is built with many interface and expansion ports, both in hardware and software, to allow for future enhancements.

Software Design Considerations

The software package needed to be designed from the beginning to have a very open architecture. This would allow for revisions and modifications to be implemented quickly. Both the test parameter and test results databases needed to be secure from tampering, but still report in a manner that would allow the manufacturers access to their own test results data. The tests would have to be administered in a separate software utility that would not be available on the distributed test systems. A test suite would have to be administered for each family of audio equipment, with the ability to customize test suites for specific models. The service center would have to report the test results to each manufacturer on a monthly basis. The system stores results in a Microsoft Access database. Drivers for the CAT.2, RF generator, and barcode printer also had to be developed.

Datappli answered the challenge and developed an automated data acquisition and control software package that a quality control (QC) operator could run and also deliver the reporting capabilities of more costly systems that usually require a higher level of experience or education to operate. National Instruments hardware and software provided Datappli with the power and flexibility to create a cost-effective solution while integrating a variety of components into one streamlined test system.

Software Design — Administration Utility

The main screen gives the user many options for administration of the test database. Each manufacturer's specifications may be administered separately from the others while maintaining the security requirements of each. All test suites may be administered as part of a family or to each model on an individual basis. Since the system is capable of running multiple automotive data buses, the administrator may select which bus to use, if any, to control the product under test. The user may add or remove tests, manufacturers, or audio equipment families from the database. Another option that we provided allows the operator to create different test suites for each manufacturer or family combination, where several different models could fall under the same test specifications. The last option allows the user to create custom test suites for any particular model that might have its own test specification. All tests can be fully customized. There is no limit to the number of tests that can be performed because they are stored in a database. The manufacturer may also determine what tests to perform on each product. As new tests or data buses are developed, they can be added to the database without the loss of previous specifications.

Software Design — Client Executable Application

The client program (Figure 1-15) is a LabVIEW executable application that is distributed with every CAT.2 system. The application controls the CAT.2 box through an RS-232 interface, runs the RF generator using a GPIB command set, and collects audio data for each test (Figure 1-16). A login screen provides a user name and password box for the operator or system administrator to enter the program. Only the system administrator has access to the Utilities Menu. The Utilities Menu provides the means for backing up databases of the results, adding users, and changing test stand parameters.

The main screen displays a summary of the test suite and the status of the tests as they progress. A new part can be scanned in using the barcode reader or entered in manually if no barcode is available. Once this is done, the program automatically calls up the database and the proper test suite is brought up for the selected part. A manual option is available for selecting only part

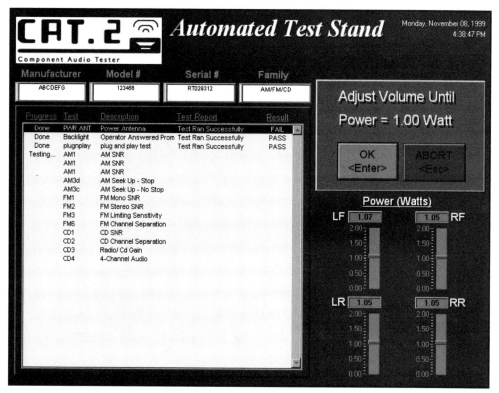

Figure 1-15
Client executable application main screen

of the test suite. Testing commences after the operator enters in a few model identification parameters and any other data collection requirements for that manufacturer. All the parameters can be fully customized to the needs of each user or manufacturer. The operator or one of the supported bus structures controls the product under test. Depending on the type of test, data is analyzed using root-mean-square (RMS), Average, and Fast Fournier Transform (FFT) functions, which are available with the LabVIEW Advanced Analysis virtual instruments. This system uses LabVIEW to do all the analysis that would have previously required many pieces of expensive test equipment. Because LabVIEW can perform complex mathematical analysis, we could replace several very expensive pieces of hardware, such as an audio analyzer and a wow and flutter meter, with software routines. All of the analysis functions have been tested and have proven to be very accurate. When the test is complete, a summary is generated on the laser-jet printer,

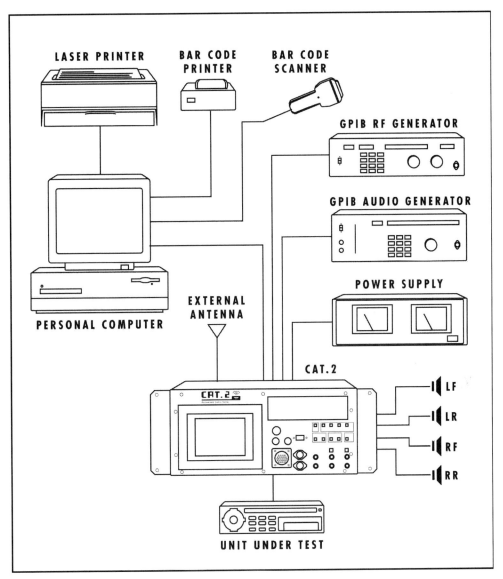

Figure 1-16
Diagram of the audio test system

the results are automatically sent to the proper SQL database, and if all tests were run and the part passed, a bar code is printed.

Design Challenges

We had to overcome many design challenges during the course of this project. The first and biggest was getting customers and manufacturers to buy in to the fact that old reliable analog test equipment can be replaced with a digital equivalent at a fraction of the cost. At the beginning of the project, there were many skeptics that thought the way that they had done it was the best and only way. As the project progressed, we saw the skeptics become advocates.

The second challenge that we faced was the need to build new test routines into the LabVIEW software. We were trying to replace analog test equipment that has been the industry standard for many years. There were many routines available to control the old equipment, but none to replace it. We had to learn how the equipment made the measurements so that we could do the same analysis in a digital realm. One example of this is the wow and flutter meter. A wow and flutter meter is used to measure the speed and low-rate frequency modulation of the audio signal. Tape speed is usually measured by playing a tape that has a recording of a known frequency and comparing that to the reproduced frequency. Tape mechanisms are a mechanical drive system that cannot ever be perfect; therefore, the instantaneous drive speed will not be constant. The user will hear the changes in this instantaneous speed as variations in pitch. Slow speed changes (<10 Hz) are called "wow" and higher speed changes (>10 Hz) are called "flutter." Instruments to measure this are very expensive, as stand-alone pieces or as part of more expensive audio analyzers. The CAT.2 is the only test system in this market that does 100% of the audio analysis in the digital domain. The audio waveforms are directly digitized by the National Instruments DAQ board and then LabVIEW analyzes all the waveforms.

The next challenge that we faced was how to keep the test time comparable with similar test systems on dedicated production lines. This challenge was met through the use of data bus control of the product under test. The system has multiple bus control devices attached to the PC. When a component model is entered, the software determines which data bus, if any, is available for that product and directs the component commands to the appropriate COM port on the PC. All of the data bus signals are fed into a single connector on the CAT.2. These control signals are passed through to the product under test using custom cables and hardware. The operator is freed from having to remember which data bus each product uses.

Another challenge was coming up with a distribution method to provide rapid field updates. This was handled by storing the test parameter and models database in an encrypted binary format. This format maintains the integrity and security of the data while keeping the file size small, allowing updates to be sent out via email.

The last major challenge was convincing the automotive electronics industry that, at a service level, parametric testing could be cost effective. Prior to the advent of this equipment, there were test solutions on the market, but none that were economical for anything other than the largest service organizations. Through the use of National Instruments products, we created a solution that is cost-effective for many of the smaller service organizations.

Results

Downtown Radio of Denver and Datappli have teamed up to successfully develop a versatile PC-based audio test system, the CAT.2. Through the use of National Instruments hardware and software, a cost-effective solution was reached; the CAT.2 provides the user with the ability to easily test any automotive audio system in a realistic manner that truly simulates the vehicle's environment. The software is intuitive and easy to learn, yet powerful enough to allow the operator to test a wide variety of audio systems from multiple manufacturers. The entire analysis portion is digital and based upon National Instruments hardware and development environment. This software, coupled with the robust and versatile hardware, provides a test system that can be applied to a wide variety of applications. The system is capable of testing the audio systems produced yesterday, the systems in production today, and the newer bus-driven digital systems that will be produced tomorrow.

■ **Contact Information**

Mark Schmitt
Project Engineer
Datappli, Inc.
3333 E. Patrick Rd.
Midland, Michigan 48642
Tel: (517) 839-1040 ext. 210
Fax: (517) 839-1042
E-mail: mschmitt@datappli.com

Ever Take a Picture of a Pothole From a Moving Truck?

Richard DeBin
Project Engineer
Datappli, Inc.

Products Used. PXI™-1000 chassis including PXI-8155B, PXI-8210, PXI-6030E, PXI-1408; SCXI-1001 chassis including SCXI-1140, SCXI-1121, SCXI-1161; ER-8 Relay Module; BNC-2110; LabVIEW 5.1; IMAQ Vision

The Challenge. Developing a system to evaluate the design and performance of containers used to transport cargo in trucks.

The Solution. Using National Instruments data acquisition (DAQ) hardware and IMAQ Vision software to create a user-friendly and cost-effective system.

Introduction

While in transit via truck (Figure 1–17), commodities can be dislodged from their location by potholes or by bumps in the road, such as when the truck crosses railroad tracks. This poses a quality liability issue for container makers like Industrial Packaging Systems (IPS). A commodity can also hit a resonance frequency in the trailer, resulting in bending or breakage. IPS, a Michigan-based packaging company, asked Datappli, Inc. to develop an integrated vehicle data acquisition (DAQ) and vision system to measure the loading inputs and the resultant forces on the cargo and the cargo containers used in truck transportation.

The goal of the project was to gain knowledge of the transportation environment, which IPS would use to make preventative engineering decisions. IPS would use the data measured by this system to improve the design and reduce the cost of cargo containers. Our challenge was to develop a system that would measure the forces exerted on cargo and the road surface that caused those forces without affecting the way the cargo was shipped. There-

Figure 1–17
Mobile test facility

fore, it was important for us to collect data such that real-world interpretations with the physics and engineering mechanics of the truck were kept in mind. The system had to be able to collect data when acceleration or strain levels exceeded a specified limit, when a pothole was detected, or when the user manually triggered the system. However, our most significant challenge was detecting potholes and capturing their images.

To date, the material handling testing services in the industry have relied on vibration tables, cam driven, single axis, and other multi-axis devices to duplicate real-world input. The results have been mixed, sometime providing false representations of the real world. IPS wanted an integrated data acquisition and vision system that would correct the deficiencies of current testing systems. These deficiencies included:

- No comprehension of the physics and engineering mechanics of truck suspensions
- Collected data not taking into account the weight of the load and suspension type
- No intuitive feel between the test and the real world.

Datappli developed an integrated data acquisition and vision system that tests the design and performance of the cargo containers and can record the cause and location of cargo damage. Why was vision included in the solution? Pardon the cliché, but a picture is worth a thousand words.

"The visual portion of the system is the tangible connection to the engineering data," said IPS President DavidWahl.

The vision element is particularly valuable in the investigative capabilities of the system. The system integrates video and engineering data to determine the origin of damage sustained by cargo. The truck system is set up to understand the cause and effect from the road profiling—photographs of the input from the road correlate to the engineering data and video from the back trailer.

The benefits of testing have already produced substantial results for IPS, which reports to date seven tests resulting in a savings of $1.6 million. In one example, a division of an automotive company wanted a replacement for a complete container system. They claimed that the container was responsible for damage to cargo in transit. The IPS Mobile Test Facility, armed with Datappli's integrated data acquisition and vision solution, proved that the current transportation method did not cause the damage. As a result, IPS avoided the total redesign and replacement of the container system.

According to Mr. Wahl, the primary benefit of the integrated system to IPS is the ability to understand the truck transportation environment. "This understanding will aid us in designing more cost effective packaging systems," he said. "The information obtained from the truck will enlighten the customer and designer to minimize over-or-under design scenarios. Packaging is a necessary evil but has great ramifications to the cost of quality. This integrated system can be utilized as a validation tool versus a predictive tool, such as vibration tables."

National Instruments (NI) hardware and software enhanced Datappli's ability to develop a seamless and quick solution for IPS. The solution, as applied to the IPS Mobile Test Facility, is easily run and monitored during a normal delivery or in a simulated transport run. To simulate a run, a normal container is used for the test. The balance of the load is simulated with hoppers full of gravel. The driver is able to control the data acquisition with just a few keyboard inputs from within the cab while driving. We selected a combination of NI's PXI and SCXI chassis to acquire the signals from all over the truck from the various types of input. By using NI hardware and software, Datappli was able to use a common platform for data measurement. The system utilizes two PXI chassis to share the tasks of measuring the loading inputs, which include road surface profiles, vehicle speed, trailer and cargo accelerations, cargo container strains, and video images from around the truck.

Hardware Design

We selected two PXI chassis for their ruggedness and located the chassis closer to the sensors to provide signal integrity. One chassis, located in the cab, measures the road surface profile and vehicle speed, records images of the road surface, and controls the video cassette recorder (VCR) that captures images of the cargo. The second PXI chassis is located under the trailer along with a SCXI chassis and modules to measure acceleration and strain and to control camera lighting.

When the sensors of the front system detect a pothole, the road surface profile is measured and a series of images of the road surface is captured. Due to the variance in the program's cycle time, this series of pictures is needed to ensure the image of the pothole is captured. A camera senses the pothole image and captures it using two PXI-1408 cards from National Instruments. Captured images from each PXI-1408 card are buffered in memory so that at the time a pothole is detected an image is transferred from a specific position in the buffer to a data file. The position in the buffer is defined by the speed of the vehicle. The front system also controls video switching of the VHS recording of cargo movement and displays the data being measured by the back system. The data from the back system is shared between the two chassis via an Ethernet connection.

The location of the back system was chosen to keep the signal wires short, which reduces the amount of noise induced on the wires. The back system is largely autonomous but can be controlled by the front system. The front system can send commands to the back system over the Ethernet connection. These commands include stopping the DAQ system, triggering a data set, and restarting the DAQ system. The ability to control other computers over an Ethernet connection allows the back system to operate without a monitor, keyboard, or mouse. This provided additional flexibility in mounting the back system. When collecting data, the back system continually monitors the sensors looking for a trigger condition. When a trigger level is exceeded, the system automatically stores a specified number of data points and then resets the hardware for the next data set.

Software Design

There were unique challenges in developing a user interface that would support the needs of IPS. The software needed to be used while the vehicle was

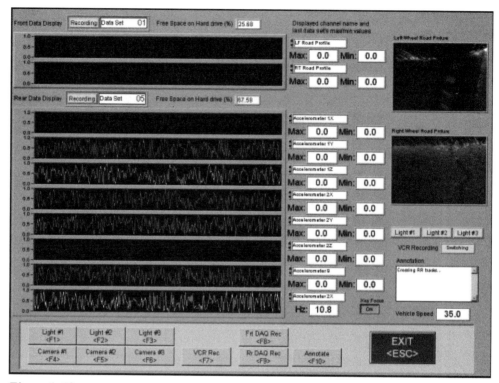

Figure 1-18
Data monitoring screen

moving, requiring it to be simple and quick to operate, yet powerful enough to control both systems from one location. LabVIEW was chosen as the preferred software editor.

The data monitoring screen, the most commonly used screen, allows the user to:

- Select which channels of data to display
- Display captured images of the road surface
- Control lighting and video recording
- Generate a text file which summarizes the test conditions
- Manually trigger data recording.

Most of the items listed above can be performed by a single keystroke on the keyboard located next to the driver.

The TCP/IP protocol compatibility of LabVIEW allowed the values of controls in the back system to be set or read from the front system; this allowed control of the program running on the back system. The TCP/IP link also allowed data collected by the back system to be read and displayed by the front system (Figure 1-18).

The remote device access (RDA) enabled the DAQ hardware connected to the back system to be controlled by Datappli's IsoCal calibration routine running on the front system over the Ethernet connection. This allowed the driver to calibrate the sensor inputs of the SCXI chassis using the system located in the cab. The newly added picture toolkit made it possible to save the captured images of the potholes in JPEG format. This picture format is efficient in its use of storage space and can be displayed on most computers.

Results

The trend towards more efficient programming, plus the higher performance of PXI DAQ equipment offered by National Instruments, made this solution possible. NI provided Datappli with the cutting-edge tools necessary to provide a superior integrated solution, resulting in a more user-friendly and cost-effective application for IPS.

■ Contact Information

Richard DeBin
Project Engineer
Datappli, Inc.
21314 Melrose Ave.
Southfield, MI 48075
Tel: (248) 353-5212 ext. 223
Fax: (248) 353-4913
E-mail: rdebin@datappli.com

Automated Radio Tester

Jeffrey Siegel, Ph.D.
Senior Project Scientist
V I Engineering, Inc.

Products Used. LabVIEW, LabVIEW Test Executive Toolkit, AT-MIO-16XE-10, SCXI-1120, SCXI-1302, PCI-GPIB.

The Challenge. Creating a system that can perform a series of tests on automobile radios that is easy to operate and requires minimal user intervention.

The Solution. Using LabVIEW and GPIB to determine the quality of automobile radios by simulating radio signals, performing audio analysis, and automating radio operations. SCXI hardware measures radio input and output voltages. Integrating the tests into the LabVIEW Test Executive Toolkit allows easy scheduling of different combinations of tests, resulting in a flexible, automated system that measures different aspects of a radio's functionality to determine if it meets quality standards.

Introduction

For a given model year, DaimlerChrysler installs in its automobiles over fifteen different radio models from three different manufacturers. To determine quality standards and to ascertain whether radios meet existing standards, DaimlerChrysler must perform over fifty different tests on each radio, measuring characteristics ranging from frequency response to interference rejection to power consumption.

V I Engineering worked with DaimlerChrysler Corporation's Radio Core Division to create a system that can perform a series of tests on a radio using various GPIB and serial instruments, data acquisition devices, and custom hardware. Using LabVIEW Virtual Instruments (VIs) to run the individual tests and customizing the LabVIEW Test Executive Toolkit to schedule the tests, V I Engineering created a system with enough flexibility to allow modification of test selection, test order, and the various parameters for each test,

Figure 1–19
Automated radio test system

while requiring a minimal amount of user intervention to run. A view of the complete system can be seen in Figure 1–19.

There are two distinct components to the automated radio tester—the tests and the scheduler. The tests control the radio and other instruments to make measurements, while the scheduler automates the running each of the tests in succession and compiles the results of each test.

Tests

Each of the tests is designed to measure and characterize a specific aspect of the radio that can be compared to pre-set limits to determine if that test passes. The tests are generally designed to measure either a single value that is characteristic of a radio, such as the time required for the seek function to traverse the band, or a series of values that are recorded while varying a single parameter, such as the power output of the radio at various audio frequencies (frequency response).

The tests are performed by simulating a radio broadcast signal from up to three radio frequency (RF) signal generators and sending the resulting signal through a simulated antenna load into the antenna input of the radio. The outputs of the radio's four audio channels (left front, right front, left rear, right rear) are sent through a simulated speaker load to an audio analyzer, which measures audio signal characteristics such as power output, distortion and signal-to-noise ratio. Other measured values include the voltage and current draw at four different power inputs to the radio, measured by SCXI-1120 modules in a chassis connected to an AT-MIO-16XE-10 data acquisition board.

Since the signal generators and audio analyzer have GPIB interfaces, most parameters of a given test can be controlled by the computer without requiring any user intervention. Parameters set with the signal generators include AM/FM, mono/stereo, RF carrier frequency and amplitude, and modulation (audio) frequency and level. A VHF switch module in a GPIB-controlled switch/control unit determines which of the RF signals is used, and a general purpose relay module determines which of five antenna loads is used. The relay module also selects the speaker load: two, four, or eight ohms. A power supply, also controlled by GPIB, sets the voltage and current to the radio.

Settings on the radio must also be adjusted during tests. Most radios for the current model year and beyond are capable of sending information and receiving commands via an automotive communication bus, either the DaimlerChrysler standard CCD or the more recent industry-wide standard, J1850. Radio controls such as the volume and the seek button can be computer-controlled with interface hardware that talks to the radio via its bus and to the computer via a serial port. This allows almost complete automated control of the entire test setup. Only changes that require mechanical adjustment on some radios, such as changing the band or tone levels, are not automated. Even for radios that do not have a bus interface, a stepper motor and a solenoid, both activated by relays in the switch/control unit, can be used to adjust the radio's volume and activate its seek button. The system allows for radios that do not have a bus interface and have an unusual faceplate configuration to be controlled manually.

Test Parameters

Although each of the tests performs a different task, they all have a common user interface and result format. Because the functionality of the tests varies widely, each test requires somewhat different input parameters to fully specify the test. For example, some tests are performed at a single user-specified audio frequency while others need multiple discrete frequencies. Others sweep the frequency, requiring a start, stop, and increment frequency. Test limits also generally have a different format for each test.

When scheduling multiple tests, it is important that all tests have the same input and output format in order to completely separate the tests from the scheduler. With a uniform format, addition, subtraction, and rearrangement of tests does not require any modification of the scheduler's code. To keep the input parameter and input type consistent for all tests, all parameters for a test are specified by a single string input. Each line of the input string contains one parameter in the format of an identifier string followed by the parameter. As an example, a line in a configuration string specifying the audio frequency would be "audio frequency (Hz) = 400". Each test parses the configuration string and extracts the values that it needs.

To completely define a test when it is run, the user must create a configuration string containing all of the parameter identifiers and values. Given that each test requires roughly 10-20 input parameters, knowing all the parameters that are needed and the exact format of the identifiers makes creating such a string a tedious task. To make the creation of the configuration string less painful, each test has a configuration program. The program has an interface screen containing controls for all the parameters relevant to the test that builds the appropriate configuration string. Figure 1-20 illustrates such a program, with the configuration string based on the controls shown in the upper right. This format string, discussed further in the next section, can be saved with a schedule configuration specifying how the test will be run.

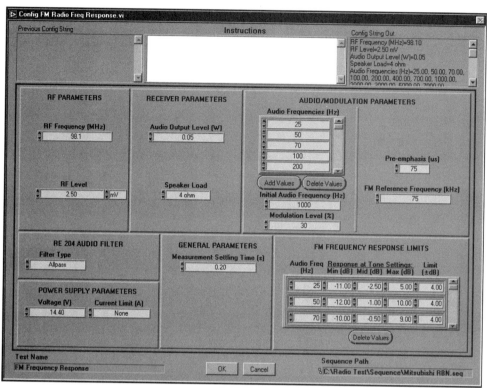

Figure 1-20
Test configuration screen

Test Results

The results of all tests are returned as a string. This string contains the relevant parameters of the test and the acquired data in a tab-delimited spreadsheet format. The test scheduler can thus receive the results of all tests, each with widely varying information, in a consistent format. This allows the test scheduler to compile all of the results and create a single result file.

In addition to presenting the test results in a format to be handled by the scheduler when a test is completed, informational screens display data as it is being collected. A status display (left panel in Figure 1-21) is displayed during every test, providing the name of the test, the test's pass/fail status,

Chapter 1 • Automotive Test

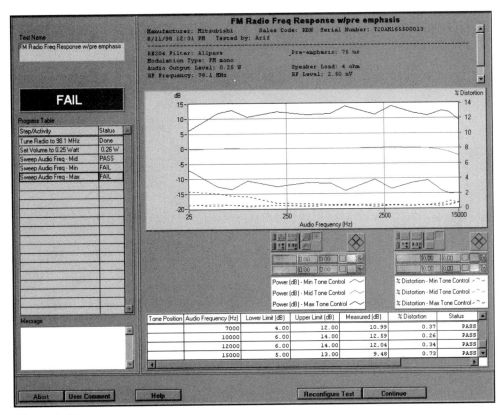

Figure 1-21
Results screen from a typical radio test

and a progress table listing the important steps within the test and the status of each one.

Results of the test are displayed in graphs, a table, or both, depending on the type of data recorded for that test. Each of the three types of displays shows the important parameters of a test, including the make and model of the radio, the date and time of the test, and the name of the user performing the test. The table displays results in a spreadsheet format, with results of one measurement in each row of the table. The graph display can present up to three overlaid graphs with independent y-axes, with each graph containing up to eight plots. A third display contains both a table and graphs, with each of the three graphs displaying up to three plots. Figure 1-21 shows this display used for the FM Frequency Response test.

Each of these displays, along with the progress display, has a very flexible programming interface that allows it to be called from within any test whenever it needs to be updated. Through this interface, the test can programmatically add test information, add a point to a plot, add a plot to a graph, and change plot names and colors at any time throughout the test.

Finally, a toolbar at the bottom of the screen is always visible during a test. The toolbar is continuously monitored during the test and contains buttons that allow the user to perform such functions as abort or pause a test, run the test again with different parameters, add comments, and get help.

Test Scheduling

Writing the code for each individual test is only half of the solution. To efficiently test a radio, a series of tests must be run in succession with minimal user intervention. National Instruments LabVIEW Test Executive Tookit perfectly meets this requirement. Given a collection of single test VIs, the Test Executive Toolkit allows a user to create savable schedules, each of which can contain a list of tests to be performed and the parameters to be set. Once a schedule has been created and all of the test parameters have been defined, the Test Executive Toolkit can execute the entire schedule and display, print, and save the results of each test without the need for an operator to run each test separately.

The user security levels built into the Test Executive Toolkit are used to limit the modification privileges of users. Only users with the highest security level have the ability to create and edit schedules, modify parameters of the tests within a schedule, and run single tests within a schedule. Users with an intermediate security level can run single tests and temporarily modify certain parameters of tests. Lowest level security users may only run previously saved schedules.

DaimlerChrysler requested the ability to schedule different tests depending on the type of radio being tested. For example, a radio that is not capable of receiving AM stereo should not be tested for AM stereo channel separation. To further automate testing, the Test Executive Toolkit was customized to automatically choose a schedule appropriate for the radio being tested. An additional setup screen allows an operator to associate schedule files with particular radios. When an operator wishes to test a radio, he or she uses a bar code reader (connected via the computer's keyboard port) to scan in the serial number on the radio. The radio's manufacturer and model can be determined from the serial number and information transmitted by the radio through its communication bus. With this information, the test executive loads in the appropriate schedule.

Results

Using this automated system, the procedure to run a typical radio test is simple once the radio is physically connected: The user logs into the program, scans the serial number bar code, and clicks on the "Test Radio" button. All of the tests in the schedule appropriate for the radio being tested are automatically run. Each test automatically adjusts the radio and the other instruments. The results of each test are displayed, printed, and saved. Modifying or creating new schedules for those with the proper log-in authority is simple, since this procedure is within the framework of the Test Executive Toolkit.

With the help of LabVIEW and the LabVIEW Test Executive Toolkit, V I Engineering was able to create a powerful yet flexible automated test system. LabVIEW's built-in tools for instrument communication, and the Test Executive Toolkit schedule editor allowed programmers to focus on test functionality. The result is an application that makes complex radio testing simple.

■ Contact Information

Jeffrey Siegel

V I Engineering, Inc.
37800 Hills Tech Drive
Farmington Hills, MI 48331
Tel: (248) 489-1200
Fax: (248) 489-1904
E-mail: jsiegel@vieng.com

Anthony J. Gorro, Andrew D. Magic, and Ali Bazoun from DaimlerChrysler played a significant role in developing test specifications for this project.

Biomedical Test

Why is Biomedical Important?

The healthcare industry continues to respond to the growing trends of managed care. Because of the monetary challenge that managed care presents, clinically useful, cost-effective technologies must be developed and utilized. However, because there is less money for biomedical to spend on costly equipment, the industry must carefully spend its money. The goal is to replace large test systems containing large amounts of hardware with more productive computer-based test and measurement systems.

To reach this goal, the biomedical industry is incorporating highly advanced technologies, such as micro-electro-mechanical systems (MEMS), into their measurement device design and testing. It is essential to not only make these systems more productive with better technology, but also to make them easy to use because researchers, scientists, and doctors are teaching themselves how to use these new technologies.

What are the Present Trends and Challenges?

- **Growing use of virtual instrumentation** — Virtual instrumentation applications have spanned nearly every industry, including biomedical. In the fields of healthcare and biomedical engineering, virtual instrumentation has empowered developers and end-users to conceive, develop, and implement a variety of research-based biomedical applications and executive information tools. Traditional use of medical equipment with stand-alone instruments has become outdated and too costly. The industry is turning to a PC-based system that includes virtual instruments.

- **Increased need for more reliable and sophisticated data monitoring systems** — Today's industry requires engineers to rapidly design more complex and more sophisticated data monitoring systems. Data monitoring of physiological responses is used to research disease, emotional stress and environments, the effects of drugs on blood pressure, and more.

    New, more stringent requirements are being placed on medical device manufacturers that must address mandates to the satisfaction of the FDA. Additionally, more component testing is being done, and advanced in-process or product testing has become the norm for the industry. Together, these efforts help ensure safe, reliable, well-designed, and well-tested products reach the marketplace.

    "Cutting Latency on Assessing Heart Period Variability Studies," shows how an updated data monitoring system is more effective and more reliable.

- **Costly equipment and instrumentation** — The biomedical industry faces serious cost controls for equipment. Traditionally, large amounts of money could be spent on medical equipment and testing equipment. With controlled costs, the demand has increased for lower-cost, more flexible instruments with more powerful processing capabilities, faster data transfer, and more efficient data storage. As medical researchers attempt to meet this demand they have found that traditional medical instrumentation is too costly to maintain for the limited capability it offers.

- **Explosion of data** — The biomedical industry is experiencing a data explosion. Because traditional systems have not been very sophisticated, researchers are looking for a way to organize and present this data with current technology. They face the need for more sophisticated technology. Researchers use computer-based measurement and automation to use the most recent hardware and software technology. Because it is very difficult to sift through information and present it in a logical manner, it is becoming evident that the best way to organize and make sense of data is to use computer-based systems.

  The paper entitled "A Cardiovascular Pressure-Dimension Analysis System" demonstrates how information can be acquired and analyzed with a computer-based system.

- **Increased use of the Internet** — Another major industry trend is an increased use of the Internet. The Web is used to measure and share data among researchers, scientists, and doctors. Almost all information obtained in the biomedical industry is shared across the Web to quickly communicate results. The industry needs systems that acquire, analyze, and present data, but also can communicate it to other sources, such as the Web.

What are the Future Trends and Challenges?

Research and development efforts are on accelerated development paths that need electrical and mechanical properties evaluated quickly. Researchers and scientists in the biomedical industry will continue looking to create PC-based monitoring systems that can run quickly and effectively at an affordable cost. It will be important for these systems to be easy-to-use, because often there is no specialist to teach the researchers, scientists, and doctors how to use these new computer-based systems.

How does National Instruments Fit In?

As application needs continue to change, virtual instrumentation systems continue to adapt. National Instruments LabVIEW is a system that can provide long-term flexibility to meet the needs of the biomedical industry. LabVIEW will not require new hardware to simply upgrade the system.

As the need to increase research, development, and production in the industry rises, LabVIEW can provide the capabilities needed to strengthen the design and test process. LabVIEW provides the necessary tools to acquire, analyze, and present sophisticated data. Again, the edge LabVIEW provides in researching and developing could mean the difference between success and failure in the highly competitive biomedical device marketplace.

Cutting Latency on Assessing Heart Period Variability Studies

Keita Ikeda
Senior Project Manager
Digital Instruments, Inc.

Dr. Bradley V. Vaughn
Department of Neurology
University of North Carolina at Chapel Hill

Dr. Stephen R. Quint
Biomedical Engineering Department
University of North Carolina at Chapel Hill

Products Used. LabVIEW, PCI-MIO-16E-4.

The Challenge. Updating a legacy system and improving user interface for equipment to assess heart period variability (HPV) studies. Traditionally, HPV study data on sleep disorder patients were collected and scored by a computer, and then manually inspected for inaccuracies. This would oftentimes mean hours of sitting in front of a console, going through and marking and verifying each QRS complex. We needed a way to quickly assess sleep studies containing up to eleven hours of electrocardiogram (ECG) data for HPV.

The Solution. Using National Instruments graphical software development package, LabVIEW, to create a suite of user-friendly programs to quickly sift through the data to mark the QRS complex and tally the HPV data and edit its results. We used LabVIEW Advanced Analysis Package to implement difficult signal processing techniques that dramatically reduced the number of false QRS detections. In our suite of programs we were able to automate the collection and calculation of the HPV data and error checking for a quick turnaround time of 26 minutes per 9-hour study.

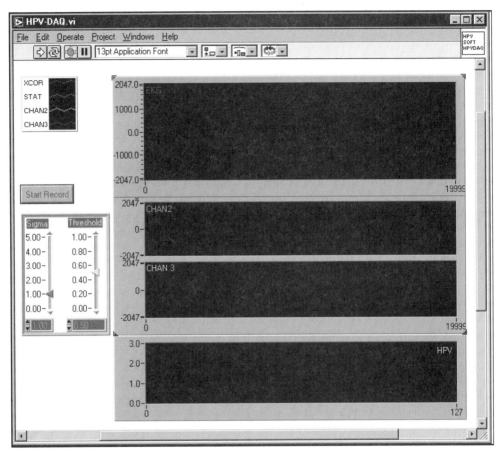

Figure 2–1
Main instrument panel for the updated HPV system

Introduction

The project began with updating a legacy system based on DIGITAL RT11, with the software system written in FORTRAN. Our objectives were to create a robust, easy-to-use and easily upgradeable system while keeping costs in check. We needed to migrate from outdated equipment that was no longer used in the industry. We chose National Instruments data acquisition (DAQ) boards and development software to ensure longevity of the system through hardware upgradeability and software flexibility with LabVIEW. The main instrument panel for the updated system is shown in Figure 2-1.

System Hardware

Usually, we would write higher-level drivers with the dynamic link library (DLL) supplied by the manufacturer of the DAQ board. We would take on the laborious task of writing routines to parse strings and serialize data. The choice of a DAQ board from National Instruments reduced our development time because the high-level drivers were already available to use in the LabVIEW environment.

To insure upgradeability and protect ourselves from obsolescence, we began with selecting a generic 450 MHz Pentium II processor PC equipped with a NI DAQ board, the PCI-MIO-16E-4. The PCI-MIO-16E-4 was used to digitize the analog signal from a GRASS instruments physiological preamplifier. We chose it because of its instrumentation-quality circuit design, open architecture system design, and ease of integration with other industry-standard components and software. The PCI MITE ASIC implementation on the PCI board allowed us to achieve greater throughput while reducing processor overhead. The liberated processor time allowed us to explore many avenues of real-time data processing in software that was previously impossible.

System Software

Instead of migrating RT11-native FORTRAN code to a PC, we started programming the software system to do data acquisition and processing from scratch. Choosing LabVIEW and its graphical programming interface allowed us to get a prototype version of the data acquisition (DAQ) software up and running within hours of opening the LabVIEW box. Within six weeks, we were able to incorporate features such as automatic study protocol setup, automatic HPV data error checking, and an analysis program that incorporates advanced signal processing. With the LabVIEW Advanced Analysis Package, we were able to incorporate a sophisticated QRS detection

Chapter 2 • Biomedical Test

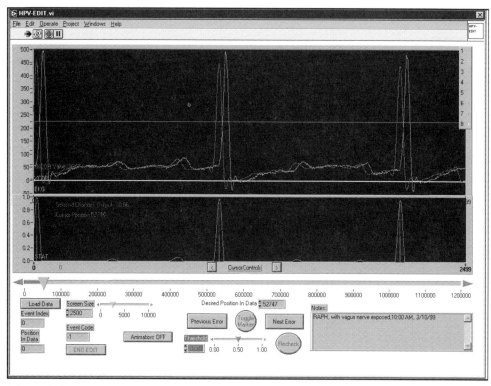

Figure 2–2
Instrument panel for the user-friendly editing and proofing software

algorithm based on cross correlation and statistics. The graphical and the instrumentation-centric nature of LabVIEW allowed us to quickly:

- Incorporate design turns
- Design the graphical user interface according to the clinician's needs
- Set up different instrumentation settings for different study protocols
- Put together a suite of complementary software that gives the clinician a wide range of tools to analyze data
- Diagnose problems within the software.

The LabVIEW programming environment allowed us to cut development time and costs while increasing the functionality and user-friendliness of the heart period variability analysis system. The instrument panel for the user-friendly editing and proofing software is shown in Figure 2–2.

Benchmark Measurements

To better understand the benefits we reaped by using National Instruments hardware and software products, we evaluated the project on three different criteria—cost of materials, functionality, and performance.

Initial Costs

The entry costs of developing the HPV analysis system (a ballpark comparison of the costs involved since the original system was developed over two decades ago).

RT11+VT100	$11,000
A/D board	$ 5,000
GRASS P511	$ 995
Total (in 1977 dollars)	**$16,995**
Current system:	
Pentium II 450 MHz PC	$ 2,240
NI PCI-MIO 16E-4	$ 995
GRASS P511	$ 995
LabVIEW FDS	$ 1,995
Total	**$ 6,225**

The original system cost 200% more, even before adjusting for inflation.

Functionality

The RT11-based system recorded the ECG trace and performed a first-order and amplitude threshold criteria filter to mark the QRS complex. It allowed the user to manually step through the data and correct false detections, and verify the HPV data. The LabVIEW/Pentium II-based system recorded the ECG and performed a cross-correlation analysis with statistical threshold filtering to mark the QRS complex. It also created a database of unusual events in the ECG/HPV data so the user could zero in on the false detections and other anomalies quickly. The graphical user interface (GUI) editing software allowed the user to quickly jump from error to error, thereby reducing the time spent manually verifying HPV data.

Performance

To quantify this criteria, we compared the number of false detections for a 9-hour overnight sleep study and the time it took for a technician to verify the HPV data (turnaround time).

False Detections:
- FORTRAN/RT11   ~270 errors
- LabVIEW/Pentium II   21 errors

Turnaround Time:
- FORTRAN/RT11   ~108 hours
- LabVIEW/Pentium II   26 minutes

Results

The benchmarks show that we have been able to leverage the increasing power of the PC with the help of National Instruments products. Using LabVIEW in conjunction with NI-DAQ boards decreased our development time and increased productivity of the sleep lab while keeping costs low. The performance increases easily dwarf the cost of ownership of the system. And the open architecture design of this system and National Instruments products lets us easily accommodate technology advances, making it difficult for it to become obsolete in the future.

■ Contact Information

Keita Ikeda

Senior Project Manager
Digital Instruments, Inc.
Mill Creek Suite E-18
708 Airport Rd.
Chapel Hill, NC 27514
Tel: (919) 942-3125
Fax: (919) 942-5797
E-mail: Ikeda@bme.unc.edu

A Cardiovascular Pressure-Dimension Analysis System

Eric Rosow, M.S.
Director of Biomedical Engineering
Hartford Hospital

Joseph Adam, M.S.
President
Premise Development Corp.

Chris Roth, B.S.
Software Engineer
Premise Development Corp.

Products Used. LabVIEW, DAQCard™-700, DAQCard-1200, and PCI-MIO E Series board.

The Challenge. Building a flexible, cost-effective, and easy-to-use research measurement system which can acquire and analyze disparate physiological signals from diagnostic ultrasound machines, electrocardiographs, and patient monitors.

The Solution. Developing an automated data acquisition and analysis system was developed which allows cardiologists and researchers to perform online and retrospective cardiovascular pressure-dimension and stroke work analyses during routine cardiac catheterizations and open-heart surgery.

Introduction

The intrinsic contractility of the heart muscle (myocardium) is the single most important determinant of prognosis in virtually all diseases affecting the heart (e.g., coronary artery disease, valvular heart disease, and cardiomyopathy). Furthermore, it is clinically important to be able to evaluate and track myocardial function in other situations, including chemotherapy (where cardiac dysfunction may be a side effect of treatment) and liver disease (where cardiac dysfunction may complicate the disease).

The most commonly used measure of cardiac performance is the ejection fraction. Although it does provide some measure of intrinsic myocardial performance, it is also heavily influenced by other factors such as heart rate and loading conditions (i.e., the amount of blood returning to the heart and the pressure against which the heart ejects blood).

Other indices of myocardial function based on the relationship between pressure and volume throughout the cardiac cycle (pressure-volume loops) exist. However, these methods have been limited because they require the ability to track ventricular volume continuously during rapidly changing loading conditions. While there are many techniques to measure volume under steady state situations, or at end-diastole and end-systole (the basis of ejection fraction determinations), few have the potential to record volume during changing loading conditions.

Echocardiography can provide online images of the heart with high temporal resolution (typically 30 frames per second). Since echocardiography is radiation-free and has no identifiable toxicity, it is ideally suited to pressure-volume analyses. Until recently however, its use for this purpose has been limited by the need for manual tracing of the endocardial borders, an extremely tedious and time-consuming endeavor.

The System

The development of an automated, online method of tracing endocardial borders (Hewlett Packard's Acoustic Quantification [AQ] Technology) has provided a method for rapid online area and volume determinations. Premise Development Corporation, in collaboration with Hartford Hospital, has developed a sophisticated application that acquires echocardiographic volume and area information (in conjunction with ventricular pressure and ECG signals) to rapidly perform pressure-volume and pressure-area analyses. The development and validation of this system has led to numerous abstracts and publications at national conferences, including the American Heart Association, the American College of Cardiology and the American Society of Echocardiography. This system is fully automated and allows cardiologists and researchers to perform online pressure-dimension and stroke work analyses during routine cardiac catheterizations and open-heart surgery.

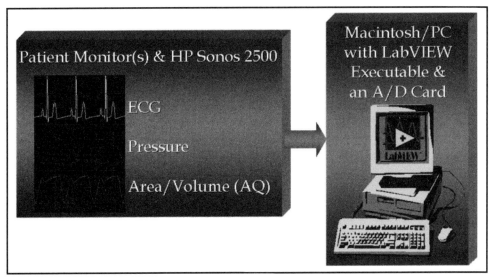

Figure 2-3
Schematic of controls/indicators

Figure 2-3 illustrates the measured parameters and the specific hardware used for this application. Although initially developed on a Macintosh platform, thanks to the cross-platform capabilities of LabVIEW, the system has been ported over to a Windows NT/98/95 environment and currently runs under LabVIEW 5.1.

Data Acquisition and Analysis

Upon launching this application, the user is presented with a dialog box that reviews the license agreement and limited warranty. Next, the main menu is displayed, allowing the user to select from one of six options as shown in Figure 2-4.

When conducting a test, an automated calibration sequence for the pressure and ultrasound signals (acoustic quantification) is generally performed. If the user elects not to perform a calibration, the most recent calibration values are accessed from the integrated calibration log. Pressure calibration data is retrieved from a "lookup table" containing specific gain and offset values for a variety of different manufacturers' pressure monitors. The AQ calibration procedure involves "scaling" and "mapping" the display image signal

Chapter 2 • Biomedical Test

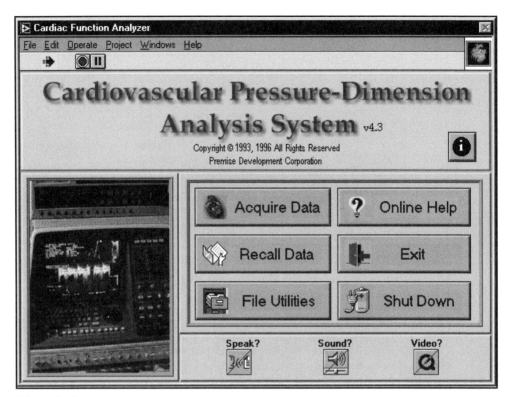

Figure 2-4
Cardiovascular Pressure-Dimension Analysis main menu

over the -1 to +1 volt output range. Figure 2-4 illustrates the front panel for the Hewlett Packard Sonos 1500/2500/5500 calibration procedure. Sequential instructions in the form of a scrolling string indicator, as well as dialog boxes, are also available.

The default sampling frequency for each channel is 200 Hz. Data is typically collected for 20 to 60 seconds. The user is presented with a pre-scan panel (as shown in Figure 2-5) to ensure that each signal is calibrated and tracking appropriately. When the user is ready to collect and store data, the Cardiac DAQ sub-VI is called. This sub-VI uses double-buffering to collect and display each channel for the pre-defined time. In order to maintain high temporal resolution, data is displayed in 10-second sweeps. An indicator is provided to display the instantaneous and total collection time. Attribute nodes of the multiplot graph are used to auto-scale each waveform and to temporarily store the data until it is written as a new data file.

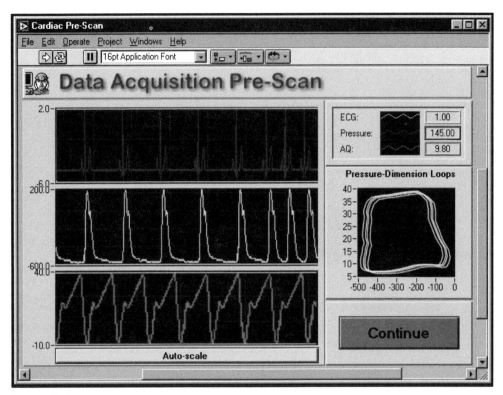

Figure 2-5
Cardiac DAQ front panel

Once data is collected, the user is presented with the data selection sub-VI to define a particular range of data to save to a file. This option allows the user to store only that portion or subset of data which is useful amongst the entire collected data set (i.e., the last 25 seconds of a 60 second array). The default setting will store the entire data set. Graph cursors are used to interactively set the initial and final indices of the data subset as illustrated below.

Figure 2-6 illustrates how an isochronic landmark can be determined for each cardiac cycle. This landmark is typically calculated from the QRS complex of the ECG waveform. However, during some studies, the ECG signal and a baseline drift may not allow for each landmark to be accurately determined. In this situation, the CPDA system allows the user to determine new isochronic points by calculating the first derivative of the pressure waveform and identifying the time indices where this signal equals zero.

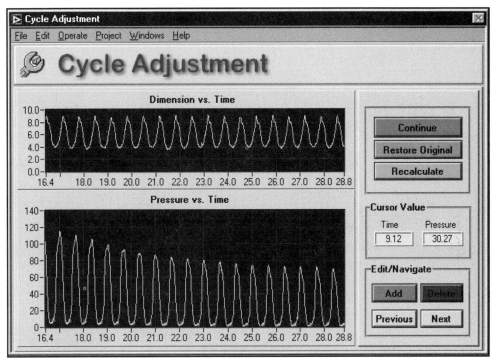

Figure 2–6
Calculation of isochronic landmarks

Time-varying elastance is measured by determining the maximum slope of a regression line through a series of isochronic pressure-volume coordinates. This parameter is determined and profiled in the Isochronic Analysis panel.

Stroke work is calculated by quantifying the area of each pressure-volume loop. Statistical parameters are also calculated and displayed for each set of data. A derivative of the stroke work calculation is a parameter called the Preload Recruitable Stroke Work Index (PRSWI) in which the stroke work is plotted against the diastolic dimension. The slope of this regression line is also a sensitive indicator of cardiac contractility.

Figure 2–7 illustrates a feature called interactive phase-delays. This feature was developed to accommodate distal (and non-invasive) pressure measurements, which need to be shifted with respect to time to yield appropriate pressure-dimension loops. Interactive graphs allow the user to shift any array of data forward or backward in time. Absolute and relative time shifts

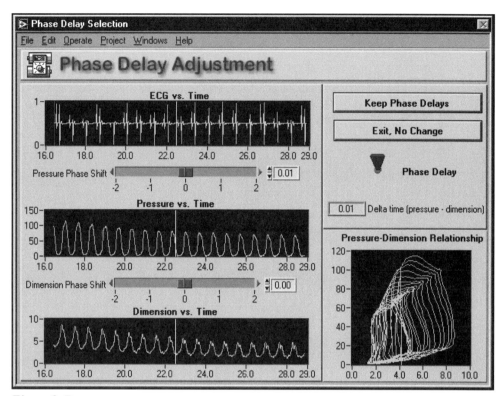

Figure 2-7
Interactive phase delay adjustment screen

are indicated along the ECG graph and the resulting pressure-dimension loops can be examined on-line.

Clinical Significance

Although no more than four channels are currently collected, several important relationships can be derived from these signals. Specifically, a parameter called the End-Systolic Pressure-Volume Relationship (ESPVR) describes the line of best fit through the peak-ratio (maximum pressure with respect to minimum volume) coordinates from a series of pressure-volume loops generated under varying loading conditions. The slope of this line has been shown to be a sensitive index of myocardial contractility that is independent of loading conditions. Several other analyses, including: time varying

Chapter 2 • Biomedical Test

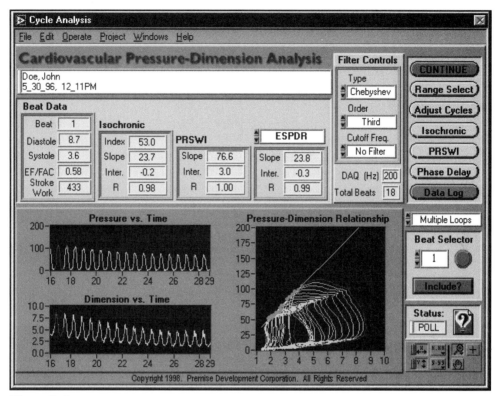

Figure 2-8
The Cardiac Cycle Analysis front panel

elastance (E_{max}) and stroke work, -dP/dt, and End-Diastolic Pressure-Dimension Ratios (EDPDR), the relaxation constant and passive stiffness constants, ejection fraction/fractional area change, and systolic and diastolic values for each beat are calculated. In addition, single loops or an entire set of loops can be highlighted in which the ECG, pressure, area/volume and corresponding pressure-dimension loop(s) are displayed in red. Finally, advanced filtering options employ Butterworth and Chebyshev filters with user-definable cutoff frequencies and filter order, and built-in statistical analyses are available on all selected data arrays and calculated parameters (i.e., R-value, ANOVAs, etc.). (See Figure 2-8.)

LabVIEW Tips and Techniques

Several features inherent in the dataflow paradigm and LabVIEW have been incorporated throughout this and other biomedical applications. Many of the interactive features are the result of a finite state machine model that was implemented using the LabVIEW case structure within a while loop. Shift registers are used for local storage. This configuration creates an "event-driven" application, which is extremely modular and easy to debug and enhance.

Global variables and configuration files are used to store site-specific data such as device number, file paths, and system parameters, while charts and graphs are used as local buffers. In addition, with respect to memory management, the preferences are set to deallocate immediately.

Significant re-design has been devoted to enhance the graphing efficiency of these applications. LabVIEW cursors greatly improve the speed at which the user can specify subsets of data, and zoom and panning controls allow substantial flexibility in manipulating graphical data. Graphing itself has been improved, in that once a set of multiplot data is calculated and displayed, subsequent replots utilize the state-machine model to recalculate only those portions of the multiplot that require recalculation.

It is essential for the user to have the ability to select and analyze subsets of data—rather than the entire dataset. This includes the ability to remove outlier data points, filter, export, and log many data sets. Datalogging and file management sub-VIs are used to accomplish these tasks. All data sets are run through a diagnostic verification to ensure data integrity when accessing from a file. Finally automated backup and import/export features allow for data arrays, calibration factors, comments, etc. to be analyzed and incorporated into third-party applications. The import feature allows for the analysis of data arrays from other measurement systems. Edited data sets can be easily logged and instantly retrieved for comparative and retrospective analysis.

Conclusion

Continued development of this system will involve further refinement of the previously described analyses to extend clinical applicability. In addition to Hartford Hospital, this system is currently being used by leading medical and research institutions throughout the world (Figure 2-9). Current efforts focus on the validation of a non-invasive means to derive indices of left and

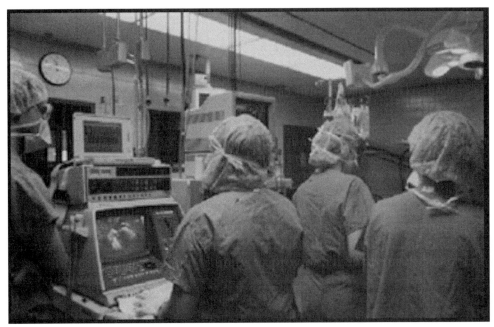

Figure 2-9
Cardiovascular Pressure Dimension Analysis System in the operating room using the DAQ-700 PCMCIA Card.

right ventricular myocardial function by eliminating the need for direct determinations of invasive ventricular pressures. If successful, it will be possible to perform analysis of myocardial function during the course of a routine echocardiographic examination.

Finally, the multi-platform capability of LabVIEW allows the system to run under several operating systems and on a variety of computers and hardware platforms. This ensures its applicability in the global healthcare arena and maximizes its opportunity to provide clinically useful data in a cost-effective manner.

■ Contact Information

Joseph Adam
President
Premise Development Corp.
36 Cambridge Street
Hartford, CT 06110
Tel: (860) 673-0484
Fax: (860) 523-9822
E-mail: joe.adam@premisedevelopment.com

PC-Based Vision System for Wound Healing Assessment

Tzu-Chien Hsiao
Institute of Biomedical Engineering
National Yang-Ming University

Huihua Kenny Chiang
Institute of Biomedical Engineering
National Yang-Ming University

Eng-Kean Yeong
Department of Clinical Surgery
National Taiwan University Hospital

Chii-Wann Lin
Institute of Biomedical Engineering
College of Medicine and College of Engineering
National Taiwan University

Products Used. LabVIEW 5.1, DAQCard-700, IMAQ PCI-1408, and IMAQ Vision.

The Challenge. Developing a flexible, PC-based image vision system for wound healing assessment. Combined with the advanced analysis module, it can be used in clinical settings to reduce the need for subsequential skin surgery.

The Solution. Using LabVIEW and IMAQ Vision to create a graphical user interface (GUI), and to perform both the programming of clinical tasks and the processing of the images with the IMAQ board. All this is done within a portable computer that reads the image from the charge-coupled device (CCD) camera.

Introduction

Efficiently developing and building a vision system for wound healing assessment is substantially important in modern burn care. Such a vision

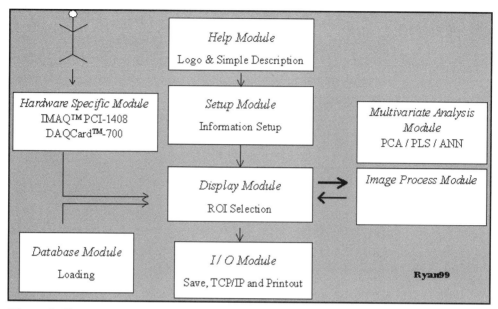

Figure 2-10
The general architecture for the PC-based vision system

system can provide critical evaluation parameters for both diagnosis and therapy. For example, the spatial and spectral color distributions on a patient's injured region are vital for a doctor to make a decision on whether to perform a skin graft.

The general protocol for the vision system has been applied in developing various virtual medical instruments [1]. Using LabVIEW running on a portable PC, we developed a virtual vision system for wound healing assessment and control of the diagnostic process. We recorded the patient's injured region by using a CCD carmera and a PCI-1408 IMAQ board. We also used a DAQCard-700 for diagnostics process control.

The system uses many functions in both LabVIEW and IMAQ Vision, including data acquisition, image acquisition, and advanced analysis techniques, e.g., multivariate analysis, artificial neural network, and matrix manipulations. Once the PC-based vision system has been adopted in clinical use, the image display module and the advanced module's image processing and multivariate analysis capabiliites can calculate properties including size, histogram, percentage, and presence of color region.

System Hardware and Software

We used Intel Corporation's MMX technology because it improves the performance of vision software and significantly accelerates integer or fixed-point arithmetic functions that are used to process images [2]. The vision system was designed and programmed on a Acer PCX-370 portable computer with an Intel-166 MMX CPU. The acquired image is transferred from a CCD camera (Sony Mode XC-77) through a PCI image acquisition card (IMAQ PCI-1408). All image acquisition and display are performed with software written in LabVIEW 5.01 and IMAQ Vision.

The general protocol for virtual medical instruments with several developed modules, illustrated in Figure 2-10, can be easily adapted to help the design of this PC-based vision system. For example, the setup module can record or retrieve patient information (name, age, ID number, and so on), environmental information (date and temperature), and instrument information. Users can browse the acquired image and present it in the display module. The general protocol also allows the operator to select the region of interest. In addition, it includes the advanced module that includes the classical statistical analysis methods [3], artificial neural network [4], and image process [5]. The general protocol also includes networking and database access.

Results

We investigated the image acquisition and analytical technology that can be applied to wound healing assessment. We were the first to integrate image processing with the diagnostic process in clinical skin surgery at National Taiwan University Hospital.

In the past, many turnkey vision systems were rendered, defined, and built for specific medical vision tasks, such as MRI, CT, PET, and endoscopy. These systems have good performance and require little time to develop and install; however, it is difficult to adapt these delicate vision systems to new applications that may arise. This type of vision system is expensive. It is not easy to improve its performance over time by upgrading the system with the latest PC or a different operating system.

An understanding of how the medical diagnostic process works and the type of signals that are acquired by a computer are the keys to designing a successful PC-based virtual medical instrument. Because our application in biomedical engineering has its own unique requirements, application software tools must provide flexibility for a wide range of solutions. The success of this PC-based vision system for wound healing assessment is largely due to its flexibility. Extensive image acquisition, spectra acquisition, advanced analysis, presentation, and networking capabilities are available within a single package, so that the operator, probably a doctor or a biomedical engineer, can seamlessly create a complete solution.

In this study, a PC-based vision system combined with data acquisition was developed to perform a wide range of inspection tasks, including using algorithms for better and more accurate diagnosis with the goal of reducing the number of painful skin surgeries that are performed. This vision system under the LabVIEW environment can be used for both online and offline applications such as non-invasive monitoring and diagnosis. Our future research will focus on finding the significant pattern for burn wounds and increasing the validity of burn wound assessment management.

References

[1] "General Protocol for Virtual Medical Instrument—the Application of Non-Invasive Blood Glucose Monitoring," *NIDAY-TAIWAN 98*, July 21, 1998.

[2] "National Instruments 1999: Measurement and Automation Catalogue", *National Instruments, Austin*, p 510-511, 1999.

[3] "Multivariate Calibration", edited by Harald Martens and Tormod Naes, *John Wiley & Sons*, p 73-164, 1996.

[4] "Neural Network: A Comprehensive Foundation", edited by Simon Haykin, *Prentice Hall*, p 156-256, 1999.

[5] "Digital Image Processing: Principles and Applications", edited by Gregory A. Baxes, *New York: Wiley*, 1994.

■ Contact Information

Dr. Chii-Wann Lin
Institute of Biomedical Engineering
College of Medicine and College of Engineering
National Taiwan University
No. 1, Jenai Road Sec. 1
Taipei 100, Taiwan
Tel: 886-2-23970800 ext. 1446
Fax: 886-2-23940049
E-mail: cwlin@cbme.mc.ntu.edu.tw

Biomedical Patient Monitoring, Data Acquisition, and Playback with LabVIEW

Guy A. Drew
Senior Electronics Engineer
U.S. Army Institute of Surgical Research

Steven C. Koenig, Ph.D.
Assistant Professor of Surgery
Jewish Hospital Heart and Lung Institute, Louisville, KY

Products Used. LabVIEW, DAQ Wizard, 16-channel A/D board AT-MIO-16E-10.

The Challenge. Providing real-time patient monitoring, data acquisition, and post-data collection review during intra-operative surgical procedures and post-operative recovery.

The Solution. Combining a data acquisition system (AT-MIO-16E-10 16-channel A/D board) with PC-based LabVIEW software to create a physician-friendly real-time patient monitoring, data acquisition, and playback interface.

Introduction

Investigators at the Jewish Hospital Heart and Lung Institute include cardiovascular surgeons, cardiologists, physiologists, and biomedical engineers. They actively conduct medical research designed to characterize new surgical techniques, test innovative cardiovascular devices, and evaluate pharmacological agents. During intraoperative procedures and post-operative recovery, physicians rely upon real-time physiologic measurements of cardiac function that include heart rate, blood pressures, and cardiac output to monitor surgical procedure outcomes. Further, investigators require accurate and precise data collection of physiological measurements of cardiac, systemic, and pulmonary function for post-processing analyses. Many of these experiments must also be conducted in compliance with Good Laboratory

Practice (GLP) for submission of study data to the Food and Drug Administration (FDA) to receive approval for clinical trials.

Design Challenge and Solution

Our challenge was to design a sophisticated, multi-feature data acquisition program that would be flexible enough to meet a broad range of experimental needs yet simple enough for nonengineers (physicians, cardiologists, and physiologists) to operate. Additionally, the data acquisition program had to meet FDA guidelines for GLP compliance. We developed a data acquisition system with the ability to measure up to 6 channels of pressure, 2 channels of flow, ECG, and 7 auxiliary inputs. Specifically, amplifier analog outputs are conditioned by a signal distribution system containing buffering, gain, offset, and low pass anti-aliasing filters, and then digitally converted using a 16-channel A/D board (National Instruments AT-MIO-16E-10). These digital data can be continuously displayed real-time on a PC monitor and/or saved in a data file for post-processing analyses or playback. Investigator(s) simply complete a worksheet (Figure 2–11) that provides the documentation required to configure the data acquisition program according to their study protocol.

In order to provide numerous data acquisition features and meet our physician-friendly requirements, the data acquisition program was configured to run under 4 modes of operation. First, a DAQ Wizard is used to assign channel names (i.e., Aortic Pressure), identify analog input channels (i.e., channel 1), and convert physical input values to physiologically equivalent units (i.e., 0-2 V = 0-200 mmHg). Next, the main data acquisition program is selected and the data recorder program profile menu screen is invoked (Figure 2–12).

Support documentation for experimental data sets is entered into the data file parameters and file header information panels. Individual channel names assigned in the DAQ Wizard can then be mapped in the DAQ channel setup panel. Other important user-selectable options include the ability to simultaneously record two separate data sets, select fixed data collection epochs or continuous data recording and/or load a previously defined program file. Finally, a data file into which experimental data will be stored must be identified. Indicators and pop-up warning menus provide real-time feedback to ensure no errors have been inadvertently made during configuration. Upon successful completion of these set-up requirements, run profile is selected.

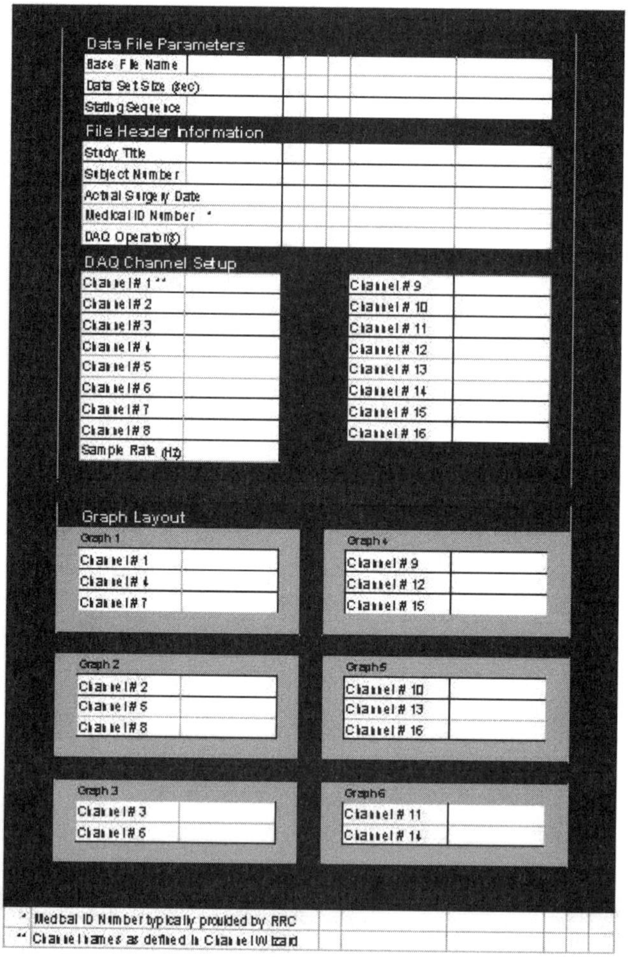
Data File Parameters
Base File Name
Data Set Size (sec)
Starting Sequence
File Header Information
Study Title
Subject Number
Actual Surgery Date
Medical ID Number *
DAQ Operator(s)
DAQ Channel Setup
Channel # 1 **
Channel # 2
Channel # 3
Channel # 4
Channel # 5
Channel # 6
Channel # 7
Channel # 8
Sample Rate (Hz)
Channel # 9
Channel # 10
Channel # 11
Channel # 12
Channel # 13
Channel # 14
Channel # 15
Channel # 16
Graph Layout
Graph 1
Channel # 1
Channel # 4
Channel # 7
Graph 4
Channel # 9
Channel # 12
Channel # 15
Graph 2
Channel # 2
Channel # 5
Channel # 8
Graph 5
Channel # 10
Channel # 13
Channel # 16
Graph 3
Channel # 3
Channel # 6
Graph 6
Channel # 11
Channel # 14
* Medical ID Number typically provided by RRC
** Channel names as defined in Channel Wizard

Figure 2–11
Worksheet user completes to configure the data acquisition system

Following run profile an automatically fitted continuous waveform chart display of up to 16 channels (one subject) or 8 channels (two subjects) is invoked (Figure 2–13). Selected patient parameters can be monitored continuously by physicians and/or stored in automatically incremented data files. Additional documentation for each data set can be logged in a notes indicator. An indicator also identifies whether continuous or epoch data function has been selected, and illuminates when invoked. A split screen feature enables simultaneous monitoring and/or data collection of separate patients.

Chapter 2 • Biomedical Test

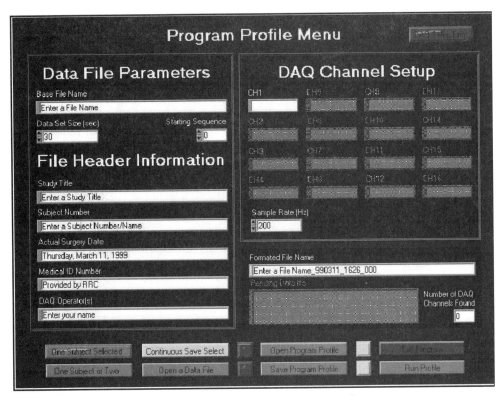

Figure 2-12
Program profile menu

There is an overlay feature that allows multiple waveforms to be displayed in the same graph (i.e., overlay aortic and left ventricular pressure). A freeze indicator allows the user to stop continuous display and study individual waveform characteristics without interrupting data collection. An exit indicator allows the user to return to the program profile menu.

A data viewer option allows the user to retrieve previously recorded data sets. The header information and ASCII data for each channel are displayed. The entire data set or individual epochs can then be replayed through data display graphs. Indicators located below the data display graphs identify starting and ending data points, and length of data set displayed. ASCII data sets can easily be imported into a spreadsheet (i.e., Microsoft Excel) or loaded in a data analysis package (i.e., HiQ™ or Matlab).

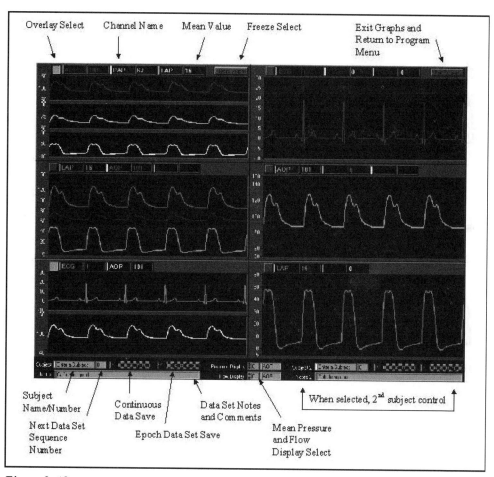

Figure 2–13
Waveform chart display

Application

Our data acquisition system and software have been successfully used in support of an on-going pre-clinical study of the AbioCor Implantable Replacement Heart (ABIOMED, Danvers, MA), per Figure 2–14. Cardiovascular surgeons, post-operative care attendants, and ABIOMED engineers

Figure 2-14
Preclinical study of the AbioCor Implantable Replacement Heart

rely on this system for subject monitoring and hourly recordings of physiological parameters for assessment of device performance and subject status.

Results

Previously, physicians and investigators relied on expensive medical-grade monitoring systems with limited ability to store digitized data and with low frequency content (< 50 Hz). Additionally, investigators relied primarily upon data chart recorders for collection of experimental data sets. These hard copy data sets limited analysis to calculation of simple cardiac performance parameters that took an extreme amount of time to complete. We selected LabVIEW as the ideal platform for developing a physician-friendly data acquisition program that could provide a low-cost alternative for real-time patient monitoring, provide more sophisticated parameter calculations for characterizing cardiovascular function, and substantially reduce post-processing data analysis time.

■ Contact Information

Guy A. Drew
Senior Electronics Engineer
U.S. Army Institute of Surgical Research
3400 Rawley E. Chambers Avenue
Building 3611
Fort Sam Houston, TX 78234-6315
Tel: (210) 916-4247
Fax: (210) 916-5992
E-mail: guy.drew@cen.amedd.army.mil

Steven C. Koenig, Ph.D.
Assistant Professor of Surgery
Jewish Hospital Heart and Lung Institute
500 South Floyd Street, Room 118
Department of Surgery
University of Louisville
Louisville, KY 40202
Tel: (502) 852-7320
Fax: (502) 852-1795
E-mail: sckoen01@athena.louisville.edu

3

SEMICONDUCTOR TEST

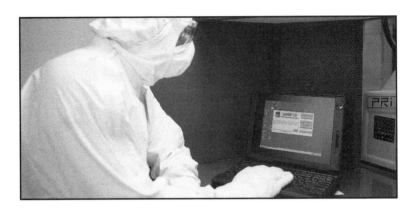

Why is Semiconductor Test Important?

Today, computers can store more files, analyze data faster, and present information more clearly than early PCs. At the same time, the size of these machines and other intelligent devices, such as hand-held computers, has decreased as their price has either remained steady or even fallen. Many of these advancements in computer technology flow directly from technological leaps in the production and test of semiconductor chips. These chips contain millions of electric circuits that form the brain of today's intelligent devices.

Manufacturing semiconductor transistors cut from semiconductor wafers in clean laboratory environments requires a complex series of steps that involve photolithography, etching, and ion implantation. Engineers use computers at each of these steps to automate the manufacturing process and to perform thorough quality tests. Because there is little room for error, these tests require software and hardware to interpret data at times on atomic scales, as detailed in the following paper, "Graphical Modeling of Quantum Atomic State Transitions in Hydrogenic Atoms."

What are the Present Trends?

As we enter the 21st century, chip manufacturers rapidly design and test semiconductor chips, knowing they must curb climbing equipment and factory automation costs. To meet the dual needs of lower costs and faster production time, engineers and scientists have steadily adopted standard computer-based technology to seize a competitive edge.

National Instruments offers low-cost computer-based tools for four areas of the $20 billion Semiconductor Industry. These areas are:

- **Device and circuit testing** — demands modular, reconfigurable high-performance tools
- **Tool automation** — can reduce development costs when using integrated computer hardware and software.
- **Imaging and inspection** — requires rapid, real-time measurements and feedback for precise imaging, inspection, and analysis of semiconductor wafers.
- **Metrology** — demands accurate and precise measurements for high-resolution wafer images.

Device and Circuit Testing

National Instruments tools provide solutions for device and circuit test applications, including parametric device characterization, analog/mixed signal characterization, dielectric and interconnect characterization, wafer-level reliability, testing of flash memory and fuse-programmable chips.

Using standard computer-based tools from National Instruments for device and circuit testing lowers the learning curve for personnel performing wafer manufacturing and tests. National Instruments currently works with leading manufacturers of wafer probers to create prober drivers for National Instruments LabVIEW, which professionals in the semiconductor industry report using more often than any other integrated test and measurement software package. Because LabVIEW integrates seamlessly with probers, instruments, and other PC-based hardware, engineers can build entire test systems for thousands of dollars less than traditional test equipment.

Factory Automation

National Instruments provides computer-based tools to automate plasma monitoring, real-time equipment control, etch monitoring, vapor deposition, sputter deposition, and mass flow controlling.

An example of factory automation using National Instruments tools is explained in the following paper, "Intelligent Automation of Electron Beam Physical Vapor Deposition." It details how LabVIEW acquires images from a CCD camera to help automate the Electron Beam Physical Vapor Deposition (EBPVD) process.

Information on a second application follows in "Data Acquisition From a Vacuummeter Controlled by RS-232 Standard Using LabVIEW." The application involves writing a LabVIEW instrument driver for a microcontroller-based vacuummeter. LabVIEW communicates with the controller across a serial connection built into the vacuummeter.

Finally, "Control System for X-Ray Photolithography Tool" explains how a control system for a prototype X-ray photolithography unit was built with LabVIEW and National Instruments data acquisition and signal conditioning hardware.

Imaging and Inspection

National Instruments tools provide customized solutions for various image acquisition and monitoring applications, including wafer imaging, defect inspection, pattern inspection, and failure analysis.

The following paper, "Angular Scanning Ellipsometer (ASELL)," details how National Instruments software and hardware work seamlessly together in an ellipsometer. A desktop PC running LabVIEW not only acquires vital data, but also controls the precise motion of mirrors that reflect polarized light from wafer samples.

Metrology

In applications such as wafer imaging, mask imaging, wafer dimensional control, and critical dimension measurement control, computer-based tools offer low-cost development tools that easily integrate with motion control and image acquisition hardware.

What are the Future Trends and Challenges?

The semiconductor industry appears poised for major growth in the coming millennium. However, during the coming years the industry will face significant technological challenges, such as lowering the cost of the design and test of complex semiconductor chips with multimillion gates, moving to larger semiconductor wafers, and using new materials in the fabrication process.

How Does National Instruments Fit In?

To meet these challenges, semiconductor manufacturers must control costs while improving overall productivity. Manufacturers can meet this challenge by using PC-based instruments, which have a significantly lower cost-of-ownership than traditional instruments. Also, manufacturers can lower costs and raise productivity by improving integration of hardware and software platforms and increasing the use of software in factory operations. National Instruments offers a wide range of cost-effective PC-based tools that manufacturers can use to meet these goals.

Angular Scanning Ellipsometer (ASELL)

Dale M. Byrne, Ph.D.
Professor
Erik Jonsson School of Engineering and Computer Science
The University of Texas at Dallas

Emmanuel M. Drège
Ph.D. Candidate
Erik Jonsson School of Engineering and Computer Science
The University of Texas at Dallas

Michael T. Kleber
B.S.E.E. Student
Erik Jonsson School of Engineering and Computer Science
The University of Texas at Dallas

Products Used. LabVIEW, NI-DAQ driver software, Lab-PC+ data acquisition board, and FlexMotion™.

The Challenge. Building a fully automated, continuous-angle, rotating-retarder ellipsometer that circumvents the need to move a large number of optical and electronic components, thus reducing mechanical complexity and subsequent systematic errors.

The Solution. Implementing an angular scanning mechanism based on a pair of stationary confocal ellipsoidal mirrors in conjunction with turning mirrors moving in tandem. The requirement for complete automation to achieve a faster and more accurate system yielded the choice of National Instruments products.

Introduction and System Requirements

Ellipsometry is a nondestructive optical technique that uses polarized light to investigate the dielectric properties of a sample. Its most common application is the analysis of very thin films. By analyzing the state of polarization of the light that is reflected from the sample, ellipsometry can yield information about layers thinner than the wavelength of the light itself, down to a single atomic layer. Depending on what is already known about the sample, the technique can probe a range of properties including the layer thickness, morphology, and/or chemical composition.

The term ellipsometry derives from the fact that the most general state of polarization is elliptic. Known for nearly a century, this very powerful and adaptable technique is primarily used in semiconductor research and fabrication to determine properties of layer stacks of thin films and interfaces between layers. Today, ellipsometry also plays a key role in many other areas, including biology and medicine, constantly posing new challenges.

The advent of high-speed digital computers has enabled scientists to develop higher quality ellipsometers. Under computer guidance, mechanical control, data acquisition, and sophisticated data analysis can be performed in a single instruments. Hence, ellipsometry is now not only an instrument sensitive technique, but also a computer intensive process.

A generic ellipsometer consists of a light source, polarizer, waveplate and focusing optics in the incident path, and recollimation optics, polarizer, and detector in the reflected path. To achieve multiple angle ellipsometry with this configuration requires varying the angle of incidence by (i) pivoting both the incident and reflected beam paths about the point of incidence on the sample, or (ii) pivoting the sample by an angle θ and either the incident or reflected path by 2θ. The importance of knowing the angle of incidence is well documented in the literature and places severe constraints on the mechanical design of the instruments, often limiting the desired angular range to a set of discrete angles.

The Angular Scanning Ellipsometer (ASELL) was developed to not only circumvent the need to move a large number of carefully aligned optical and electronic components, thus reducing mechanical complexity, but also to allow the continuous variation of the angle of incidence. This last feature results in the opportunity for more detailed investigations of the sample, as well as allows differential measurements, thereby introducing novel experimental quantities that enhance ellipsometric sensitivity to material properties. Finally the need for a fast, accurate, and user-friendly instrument required a fully automated, computer-controlled design. All those goals could not have been achieved without the products from National Instruments. It provided us with a set of integrated tools, both hardware and software, at all stages: motion control with FlexMotion, data acquisition with Lab-PC+ and NI-DAQ, and data analysis with LabVIEW.

Chapter 3 • Semiconductor Test

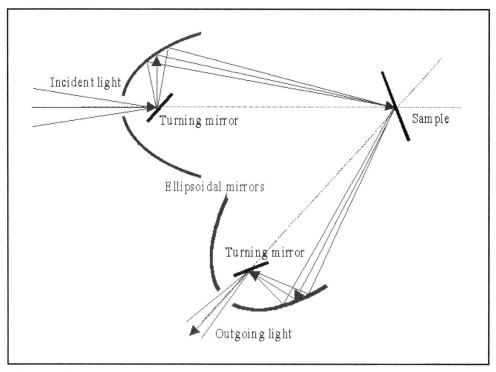

Figure 3-1
Geometry of the angular scanning mechanism

System Design

The key optical concept in the ASELL is the use of a pair of stationary confocal ellipsoidal mirrors in conjunction with two turning flat mirrors that move in tandem in order to achieve a variation in angle of incidence, as shown in Figure 3-1.

The overall layout of the ASELL prototype is visible in Figure 3-2 and comprises the following subsystems:

- **Focusing optics** — includes a monochromatic light source, polarizing components, a spatial filter, and a focusing-lens assembly.

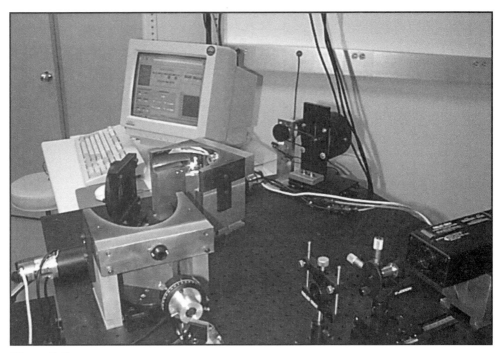

Figure 3–2
The ASELL system

- **Collimating optics** — contains a recollimating-lens assembly, a rotating-retarder arrangement, and a detector system.
- **Angular scanning optics** — includes two ellipsoidal and two flat mirrors.
- **Motion control system** — includes three rotation stages driven by micro-step motors and a multi-axis indexer.
- **Computer control system** — monitors and coordinates overall functioning, controls the motions of the stages and rotating retarder, and acquire and process data.
- **Mechanical mounts.**

From the system standpoint, the operation of the ASELL may be broken into three distinct processes: motion control, data acquisition, and data analysis. A critical aspect was for the system to be PC-based, using a software that would integrate the different tasks. While simplifying the complexity of program structure, it would also eliminate the need for interfacing between programs. LabVIEW was the ideal choice.

The motion control system is required to accomplish two tasks: the rotation of the turning mirrors and sample stage, and the rotation of a waveplate (the retarder) critical in the analysis of data. Hence it must be capable of simultaneously controlling four axes of motion via a computer. To take full advantage of the novel angular scanning concept, we needed a high-resolution system. The motion hardware chosen to accomplish the above mentioned tasks is composed of the following items:

- **A three-axis microstepping motor/drive system** — We acquired the system from Daedal/Parker to be used in the scanning mechanism. The 200-step hybrid motors have a continuous travel range (360 degrees), with a turntable gearing ratio of 36:1. A precision worm-gear provides accurate point-to-point angular positioning with minimal backlash to assure bi-directional accuracy and repeatability. Each stage is also fitted with a magnetic reed-type home switch that provides an absolute reference point to which the table always returns when instructed to do so. The motor drives result in a resolution of 20,000 steps/revolution on all three axes. To assure more sensitive control, we coupled the drives with an optical encoder, yielding a potential angular resolution of 0.0025 degrees for each motor table.

- **A 400-step hybrid step motor** — Connected to the retarder via an o-ring in a 1:1 gear ratio, it provides us with rotation speeds of up to 1800 rpm (30 revs/sec). This speed is essential for fast data acquisition. Because the position of the retarder is also critical, a micro-switch is mounted onto the arrangement and is triggered mechanically by a cam as the retarder rotates.

- **A motion controller** — An AT-6400 indexer was purchased from Compumotor to provide digital control simultaneously to all four motors.

The complexity of the motion control was greatly simplified by the use of FlexMotion. This library of virtual instruments (VIs) that runs with LabVIEW performs all the necessary motion configuration and ensures that the motion control perfectly interfaces with the rest of system.

As the retarder rotates, the intensity of the optical beam reflected from the sample varies. A photodetector is used to detect this variation in optical power and provide the user with an output voltage proportional to the intensity. The Lab-PC+ data acquisition board then samples and digitizes this analog voltage signal Also, as mentioned earlier, the retarder, as it rotates, momentarily closes the micro- switch at a particular orientation. This sends a trigger to the data acquisition board that instantly initiates the data taking process. Based on the velocity of the retarder rotation, data is acquired over a time interval that corresponds to a certain angular travel. Timing is therefore of prime importance. We opted for the Lab-PC+ as our acquisition board. It is a low-cost, multifunction analog, digital, and timing board that fits into the expansion slot of a personal computer. The Lab-PC+ is a National Instruments product. We chose it because it is interfaceable with LabVIEW, and it fulfills all requirements for the ASELL. Furthermore, the board is available with a package of data acquisition VIs that perform all low- and high-level tasks required.

The data acquired via the Lab-PC+ are processed and analyzed in order to compute the ellipsometric parameters (known as Ψ and Δ) necessary to derive the desired sample properties. The signal is first high-pass filtered to remove noise picked up from the environment and, if requested, averaged over a predetermined number of retarder cycles to eliminate stray effect. The program then performs a Fast Fourier Transform (FFT) to extract the Stokes parameters, quantities that indicate the polarized nature of the reflected beam. The data acquisition scheme is repeated for all angles of incidence. The time required for the entire process is dependent on the rate of acquisition, number of data points collected, and the velocity of the motors, all of which are accounted for by the program. LabVIEW provided us with most of the functions needed for data analysis and allowed us to easily implement within the program more specific and computationally intensive routines via its interface with Matlab.

Figure 3–3 shows the front panel of the main program. Its user-friendly interface allows users to enter the control parameters and new results viewed. An optimal configuration is preset for fast use, and complete online help is available for reconfiguration of the parameters, such as the scanning intervals, sample tilts, and scanning resolution as well as velocity, acceleration, deceleration, and starting positions for the motors.

Results

The entire system is currently being tested at the University of Texas at Dallas. ASELL is in a calibration phase that has easily justified the efforts put

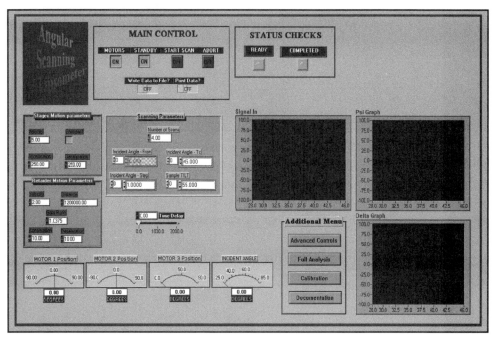

Figure 3-3
Front panel where users enter control parameters and view results

into the development. The facilitation in variation of angle of incidence, eliminating the need to move bulky optics is a prime advantage. In addition, very small angular increments (practically down to 1/300 degrees compared to the 5 degrees in the standard instruments in use in our laboratories) provide greater information about the sample, such as the exact location of the Brewster angle. It also enables the use of new differential quantities that enhance ellipsometric sensitivity. Further, the use of a focused beam on the sample results in small spotsize, maintained constant via appropriate control of the scanning mechanism, and hence allows precise localized measurements. Finally, the complete automation under LabVIEW permits accurate and rapid data acquisition and processing; only a few minutes are necessary for a high angular resolution scan of a sample.

The integration of all processes (motion control, data acquisition, and data analysis) offered by National Instruments products was the key factor in the success of ASELL. In addition, there is no need for a large computing system (the host computer is a 100 MHz Pentium with a total of 48 MB RAM).

We hope to have this product available for commercialization soon. We also plan to enhance the capabilities of the instrument by taking full advantage of the novel design and advanced features of the hardware/software setup we are using.

■ Contact Information

Emmanuel Drége
        Ph.D. Candidate
        School of Engineering and Computer Science
        University of Texas at Dallas
        P.O. Box 830688, MS:EC33
        Richardson, TX 75083-0688
        Tel: (972) 883-2979
        Fax: (972) 883-6839
        E-mail: drege@utdallas.edu

Graphical Modeling of Quantum Atomic State Transitions in Hydrogenic Atoms

Rus Belikov
Department of Electrical Engineering
Stanford University

Products Used. LabVIEW 5.1 Graph3D Control v1.0.

The Challenge. Developing a way to visualize the oscillations of atomic electron charge densities when they make state transitions under an applied electric field, e.g., in a laser.

The Solution. Using LabVIEW with its new 3D graph control, as well as the intensity graph to simulate, plot, and animate the above electron wavefunctions.

Introduction

Electron behavior under an applied electric field is at the very basis of all laser operation, not to mention many other atomic phenomena. The exact way to treat this problem is by using time-dependant perturbation theory in quantum mechanics [2], [3]. It is interesting, however, that the Classical Electron Oscillator model [1] gives results that are close to the quantum mechanical results. The reason for this is the following: The classical theory treats electrons as solid masses attached to the nucleus by a spring (as opposed to the electron clouds of quantum mechanics). An electron, excited by an external electric field of frequency ω will therefore oscillate around the nucleus with the same frequency and can be modeled as an oscillating dipole. As any oscillator, it will then have the strongest response at some resonance frequency ω_a and die off the further you get from this frequency. As it turns out, the quantum theory treatment yields a similar result. An electron wave function under an applied field ω will get partially excited or relaxed into a state with energy difference $\pm \hbar \omega_{ij}$, again with the strongest response being when $\omega = \omega_{ij}$. As we will briefly outline below, this excitation results in a wave function that oscillates with frequency ω_{ij}, thereby generating a quantum version of an electric dipole.

However, the mathematical treatments only give you an abstract view of what is going on in the atoms. To get a much more physical view, we need to graph the electron charge densities and watch their progress in time, which is exactly what I did. Atomic physics and chemistry textbooks often have diagrams of the stationary atomic states but very rarely an oscillating one, and their depictions are sometimes wrong because the computing power and programming effort required to graph something like this in the past was quite significant. LabVIEW has certain features that greatly simplify this job. It has allowed me to graphically illustrate exactly what happens to the wave functions undergoing state transitions and illustrate connections between the classical and the quantum theories.

Wave Functions Under an Applied Field

The wave function $\Psi_i(\mathbf{r},t)$ of an electron that occupies any particular energy level of an atom E_i will be an eigenfunction, or a stationary state of the Hamiltonian for that atom. What this means is that the time development of this function will be given by Shrödinger's equation as

$$\Psi_i(\mathbf{r},t) = \psi_i(\mathbf{r})e^{-iE_i t/\hbar} \qquad (1)$$

so that the probability density of position $|\Psi_i(\mathbf{r},t)|^2$ is independent of time, i.e., it is stationary.

Consider now two atomic energy levels i and j with transition frequency $\omega_{ij} = (E_j - E_i)/\hbar$. Under an applied field of frequency ω, there will be stimulated emissions between these levels, with the transition probability peaking when $\omega = \omega_{ij}$. (There will also be spontaneous emission from the higher level to the lower that is independent of the field.) The quantum mechanical reason for these emissions can be summarized as follows: Whenever there is an applied field present, there is a correction to the free atomic Hamiltonian. Thus, the Shrödinger's equation, which governs the time development of the wave function, will have to include this new Hamiltonian. As a result, the wave function will evolve in time by a different formula than (1). In fact, when $\omega = \omega_{ij}$, according to time-dependant perturbation theory [2], a good approximation is that a wave function would evolve in time as

$$\Psi_{ij}(\mathbf{r},t) \approx c_i(t)\psi_i(\mathbf{r})e^{-iE_i t/\hbar} + c_j(t)\psi_j(\mathbf{r})e^{-iE_j t/\hbar} \qquad (2)$$

for some c_i, c_j, where $c_i^2 + c_j^2 = 1$ for all t. If the field was applied at $t = 0$ to an electron initially in state i, then we have initially $c_i(0) = 1$ and $c_j(0) = 0$, and these values will evolve in time until they reach some equilibrium values $c_i(t) \to a_i$ and $c_j(t) \to a_j$. These values can be calculated by rate equation analysis [1].

Having a wavefunction, we can calculate the charge density distribution of an electron, given by $\rho(\mathbf{r},t) = -e|\Psi(\mathbf{r},t)|^2$. If we omit $-e$ for simplicity, we get from (2) at steady state

$$\begin{aligned}\rho(\mathbf{r},t) &= |a_i|^2|\psi_i(\mathbf{r})|^2 + |a_j|^2|\psi_j(\mathbf{r})|^2 \\ &+ a_1 a_j^* \psi_i(\mathbf{r})\psi_j^*(\mathbf{r})e^{i\omega_{ij}} \\ &+ a_1^* a_j \psi_i^*(\mathbf{r})\psi_j(\mathbf{r})e^{-i\omega_{ij}}\end{aligned} \qquad (3)$$

Note that the first two terms are independent of t, while the last two have a time dependence that is periodic with frequency ω_{ij}. It is precisely because of this periodicity of charge density that oscillating dipoles arise from quantum atomic transitions.

Separation of Variables in Wave Functions

For Hamiltonians that are rotation invariant and, in particular, for a simplified model of hydrogenic (single-electron) atom [2],[3] with no spin-orbit coupling, the eigenstates of the Hamiltonian can be split into radial and spherical components as follows:

$$\Psi_{nlm}(\mathbf{r}) = R_{nl}(r)Y_{lm}(\theta,\phi) \qquad (4)$$

where $\Psi_{nlm}(\mathbf{r})$ are also eigenfunctions of angular momentum squared $\mathbf{L}^2$ and angular momentum in z direction L_z. The quantum numbers n, l, m are used to index the eigenstates, with $n=1,2,3,\ldots$ corresponding to increasing energy levels $E_n \propto -n^{-2}$; $l=0,1,\ldots,n-1$, (often denoted as s, p, d,…) corresponds to $\mathbf{L}^2$ eigenvalue $l(l+1)\hbar^2$; and $m=-l,\ldots,0,\ldots,l$ corresponds to L_z eigenvalue $m\hbar$. The atomic numbers l and m simply count the degeneracies of each energy level. $R_{nl}(r)$ is called the radial wave function, while $Y_{lm}(\theta,\phi)$ is termed the spherical harmonic. Even though a lot of materials, in particular crystals, cannot be accurately modeled by hydrogenic atoms, this picture still gives very good descriptions of the qualitative behavior for such atoms.

Graphical Representation of State Transitions

A problem with plotting the entire charge density of a wave function is that we need to represent a function of three variables. The best way to do this would be to map a value of the charge density to a transparency of a pixel in 3D, with the maximal charge density corresponding to completely opaque, and 0 charge density corresponding to completely transparent. This would be the exact representation of the so-called electron cloud. However, the current computing power available is probably not enough to do animated electron clouds unless you go to very low resolution or employ some highly customized efficient algorithms.

There are alternative representations, however. An obvious one is to plot a 2D slice through the full charge density and colormap the values, which saves a considerable amount of computation and does not require 3D rendering. The two slices considered here are the x-z plane ($\phi=0,\pi$), and the x-y plane ($\theta=\pi/2$). These representations often suffice for distributions that exhibit certain key symmetries.

To get a 3D feel for what a charge density looks like however, we can do the following. Instead of taking a 2D slice through the charge density, take a spherical slice, i.e., consider (2) for a fixed r:

$$\begin{aligned}\Psi_{n_1 l_1 m_1, n_2 l_2 m_2}(\mathbf{r},t) &= a_1 R_{n_1 l_1}(r) Y_{l_1 m_1}(\theta,\phi) e^{-iE_{n_1} t/\hbar} \\ &+ a_2 R_{n_2 l_2}(r) Y_{l_2 m_2}(\theta,\phi) e^{-iE_{n_2} t/\hbar} \\ &= b_1 Y_{l_1 m_1}(\theta,\phi) e^{-iE_{n_1} t/\hbar} + b_2 Y_{l_2 m_2}(\theta,\phi) e^{-iE_{n_2} t/\hbar}\end{aligned} \quad (5)$$

where b_1, b_2 are independent of θ and ϕ, so that the magnitude squared of (5) could be computed without having to compute the entire wave function, and plotted as a surface in spherical coordinates. This will not be exactly the spherical harmonic of the full wave function in the sense of (4), of course, because $R_{n_1 l_1}(r) \neq R_{n_2 l_2}(r)$ in general, and therefore we cannot separate variables. However, all the radial wave functions have the same qualitative shape of being concentrated near the nucleus and dying off gradually, so that a plot of a spherical slice through the charge density still gives a good qualitative picture of the 3D behavior for the full charge cloud.

LabVIEW Implementation

Two LabVIEW virtual instruments (VIs) compute the wave functions according to (2) and (5) under the conditions outlined above, and graph them as shown in the figures that follow below. One of them graphs the plane slice through the probability density, while the other plots the spherical harmonic, as described earlier. Both of the programs animate the graphs so that all the details of the wave function oscillations could be clearly seen. The user interface has buttons to choose a pair of any hydrogenic wave functions up to and including $n=3$, as well as a slide that allows varying the relative strengths of the two wave functions [i.e., coefficients c_1^2, c_2^2, b_1^2, b_2^2 in (2) and (5)], among other less important controls.

Using LabVIEW has four main advantages over any other tool I can think of that I could have used for this problem.

- LabVIEW has a very rich 3D graph that allows me to plot 3D renderings of spherical harmonics in perspective, with light sources, and lets me interactively rotate and zoom in/out so I can look at the plots from any angle and position, even from the inside.

- The interactive nature of the user interface allows me to easily modify different parameters of the system simply by adjusting slides. I can pick from among several preprogrammed atomic states, continuously vary the contribution of each state as well as the speed of the oscillation, and see the change take effect in real time while the system is running.

- The 2D evaluation library within the mathematics function palette allows me to create numerical data for the plots simply by wiring in the analytical forms of wavefunctions in string format. This makes building a database of analytical wavefunctions very simple — just store them as an array of strings.

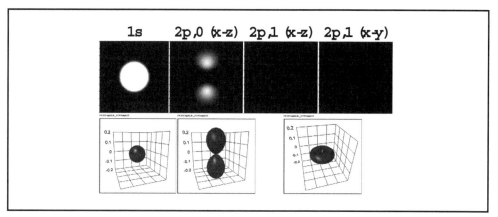

Figure 3-4
Eigenstates of the hydrogenic atom

- The data flow block-diagram nature of LabVIEW programming makes it simple to create plots and animate them — all it effectively takes is a continuous loop around the plot-generating code that advances the time parameter with each iteration.

Some of the Plots Produced by LabVIEW

The simplest possible state transitions in atoms are $1s \leftrightarrow 2p$ transitions. The $2p$ state is triply degenerate, and all of the stationary states are shown below in Figure 3-4. The quantum numbers (n,l,m) of the states shown are (1,0,0), (2,1,0), (2,1,1) respectively, with the $2p,1$ state, or (2,1,1) being shown in two planes. The remaining (2,1,-1) state looks exactly like the (2,1,1) state except that it has opposite complex phase variation. The scale of the charge density graphs is 10 atomic lengths across.

Some frames of the oscillations caused by the $1s \leftrightarrow 2p$ transition are shown in Figure 3-5 and Figure 3-6, where the contributions from the two eigenfunctions were taken to be equal, i.e., $c_1^2 = c_2^2 = b_1^2 = b_2^2 = 1/2$ in equations (2) and (5). We can see that the wave functions resemble oscillating dipoles quite closely, with linear polarization in Figure 3-5 and circular polarization in Figure 3-6. These graphs show very clearly why the electron transitions under an applied electric field are often modeled as classical electron oscillators. Another example of a more complicated transition is shown

Chapter 3 • Semiconductor Test

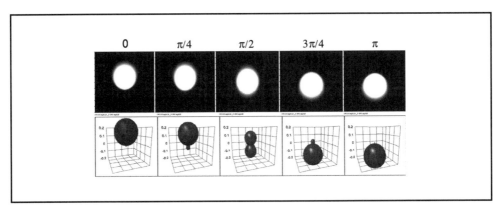

Figure 3-5
1s ↔ 2p,0 Transition oscillations

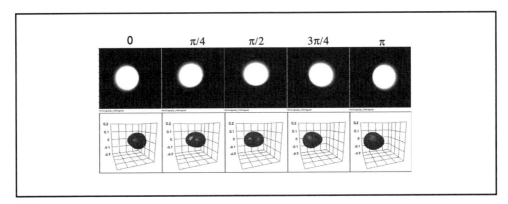

Figure 3-6
1s ↔ 2p,1 Transition oscillations (top row is view from above, i.e., the x-y plane)

in Figure 3-7 (where the scale is now 2.4 times larger across than in Figures 3-5 and 3-6). This one exhibits circular oscillations just as the transition in Figure 3-6, i.e., it is circularly polarized.

There are, however, certain transitions that do not form dipolar oscillations. As quantum mechanics shows [2], [3], these transitions are extremely rare compared to the dipolar transitions and therefore are termed forbidden transitions. They form electric quadrupoles and other higher-order electric and magnetic multipoles, and therefore cannot be modeled by the classical electron oscillator theory. Some of the examples of such transitions are shown in Figures 3-8 and 3-9 (same scale as in Figure 3-7). It can clearly be

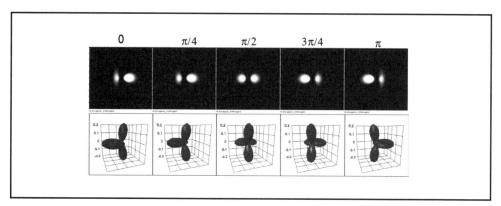

Figure 3-7
2p, 1 ↔ 3s Transition oscillations

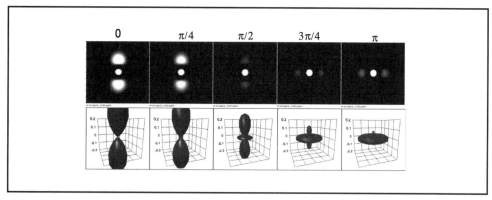

Figure 3-8
2s ↔ 3d,0 Transition oscillations

seen that there are no dipole oscillations in this case because of the symmetry. In Figure 3-9, for example, we see a rotating quadrupole.

Because of the rarity of such transitions, however, we can neglect them, and we will have the result that practically any atomic transition satisfying the conditions of our simple model will behave as an oscillating dipole at frequency given by (3), and therefore can be modeled quite well by the Classical Electron Oscillator model.

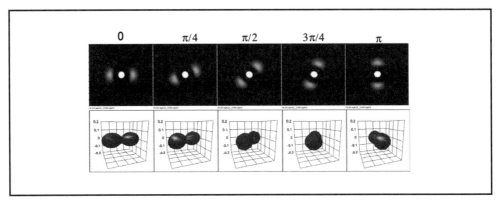

Figure 3-9
2s ↔ 3d,2 Transition oscillations (top row is in the x-y plane)

Results

Two VIs were written in LabVIEW that graphically represent the quantum atomic electron states undergoing transitions in an applied electric field. This is probably as close as theoretically possible in visualizing what electrons look like on the microscopic level. These animated graphs greatly facilitate qualitative analysis of the behavior of such transitions and in particular illuminate some of their important properties, such as a tendency to generate dipolar oscillations. A connection between the quantum and classical treatment of the subject can be made clear based on these observations, which justifies the use of the classical theory in applications such as lasers. Even though some simplifying assumptions render the analytical model used here inaccurate in some cases, the general behavior of state transitions is captured well. Certain key features of LabVIEW make it a perfect tool for this type of theoretical analysis.

References

[1] Anthony E. Siegman, *Lasers*, Sausalito, California: University Science Books, 1986, chapters 2, 3 4.

[2] B.H. Brandsen and C.J. Joachain, *Introduction to Quantum Mechanics*, Essex CM20 2JE, UK: Longman Scientific & Technical, 1995, chapters 7, 8, 11.

[3] Stephen Gaziorowicz, *Quantum Physics,* second edition, New York: John Wiley & Sons, Inc., 1996, chapters 11, 12.

■ Contact Information

Ruslan Belikov

Department of Electrical Engineering
Stanford University
144 Escondido Rd. #C
Stanford, CA 94305
Tel: (650) 497-7846
E-mail: rbelikov@stanford.edu

Intelligent Automation of Electron Beam Physical Vapor Deposition

Dr. Vittal Prabhu
Assistant Professor
Penn State University

Indraneel Fuke
Penn State University

Products Used. National Instruments IMAQ board, LabVIEW Vision and Image Processing Software, and SCXI thermocouple module.

The Challenge. Automating various operations of Electron Beam Physical Vapor Deposition (EBPVD), a coating process. The long-term objective is to automatically control the thickness and composition of coating and monitoring the process in real time. Real-time feedback of the melt pool conditions and work-piece temperature is necessary for proper control of these two parameters.

The Solution. Using SCXI thermocouple module for temperature monitoring and the Vision and Image processing software for LabVIEW for continuous analysis of the melt pool images obtained from a charge-coupled device (CCD) camera. Based on this feedback the work-piece is rotated and translated in the coating chamber to achieve optimum composition and coating thickness on the substrate component.

Introduction

Electron Beam Physical Vapor Deposition (EBPVD) is a material coating technology whereby a coating (metal, alloy, or ceramic) is melted, vaporized in a vacuum, and then deposited on a component or work-piece requiring the surface properties inherent in the coating (Figure 3-10). A strong mechanical bond, uniform microstructure, and relatively high deposition rates make it an attractive and versatile coating process. Since it is performed in a vac-

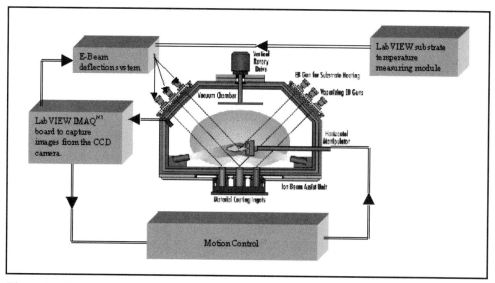

Figure 3-10
Schematic setup for EBPVD chamber and block diagrams for controlling various parameters

uum, it is an environmentally friendly technology, suitable as a replacement for other coating processes in many applications. This technique is capable of producing coatings for a wide range of industrial applications. However, at present the process is highly dependent on highly trained operators and does not possess the automatic control capabilities that would make it attractive and cost effective in manufacturing. So the ultimate goal is to have the capability to autonomously control the thickness of the coating and the rate of deposition in a real time environment. It is necessary to integrate different aspects of process control and human-machine interface to ensure good overall performance.

Monitoring and Analysis

The work-piece has to be maintained at a specific temperature to ensure good adhesion of the evaporated particles. Out of the six electron beams, two are directed on graphite plates adjacent to the substrate. The substrate is thus indirectly heated through the graphite plates. The thermocouples are mounted on the graphite plates in a mesh of 5 x 5. Continuous monitoring of

Chapter 3 • Semiconductor Test

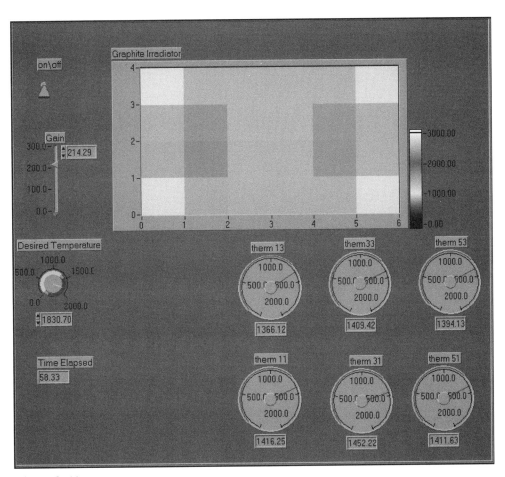

Figure 3-11
Screen capture for thermocouple module

these thermocouples using the LabVIEW thermocouple module is done, and this data acts as a feedback to the electron beam deflection system that deflects the electron beam to the under-heated part on the graphite plate. This arrangement guarantees that the component is uniformly heated to a correct temperature. The LabVIEW virtual instrument (VI) for this is shown in Figure 3-11.

The remaining four electron beams are used for evaporation. The CCD camera captures the images of melt pools of these ingots. This image is acquired via an IMAQ board. Using the Vision and Image processing software in LabVIEW, this image is compared with a known good image of a

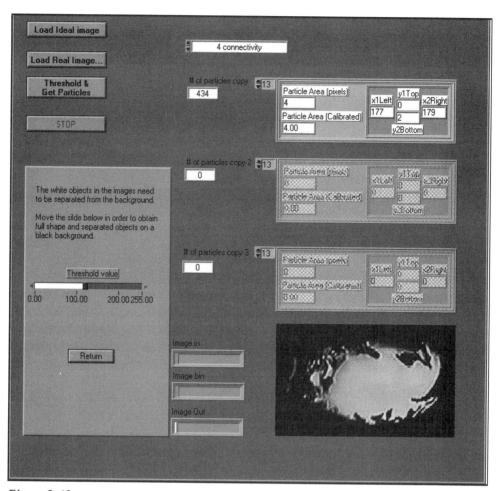

Figure 3-12
Screen capture for the image processing module

melt pool and the under-heated areas in the given melt pool are identified in Figure 3–12. This data is then fed to an electron beam deflection system and the motion controller. This allows the work-piece motion in the vapor plume to be adjusted based on the shape and intensity of the evaporated plume and the melt pool conditions in real time.

Results

EBPVD is a relatively new material coating method and it gives very good results. It is particularly useful in thermal barrier coatings, in aircraft turbine blades, and in protective coatings on cutting tools. The increased deposition rates and the variety of materials and techniques of the EBPVD process provides an opportunity to broaden the application of this technology into the U.S. marketplace and to expand its impact on military and commercial systems. EBPVD is also a member of a family of coating technologies that can replace environmentally incompatible coating processes used today. EBPVD has the advantage of working with a wide variety of materials and techniques to improve performance and increase the service life of coated components.

Industry has independently estimated the domestic economic potential of EBPVD technology over the next decade to be over $10 billion along with the creation of an additional 10,000 skilled jobs. However, a major hindrance in the process integration and control of EBPVD is that accurate comprehensive *in situ* measurements of the process parameters are often difficult and sometimes impossible because of the complexity of the process. In most cases specially trained workers with experience are needed to control the operation. This work in process control is expected to make an important contribution towards EBPVD process automation.

■ Contact Information

Vittal Prabhu, Ph.D.
Assistant Professor
Penn State University
207 Hammond
University Park, PA 16802
Tel: (814) 863-3212
Fax: (814) 863-4745
E-mail: prabhu@engr.psu.edu

Control System for X-Ray Photolithography Tool

Sergey Liberman
Principal Consultant
Solidus Integration, Inc.

Products Used. LabVIEW, data acquisition (DAQ), signal conditioning (SCXI), FieldPoint™, Industrial Automation Servers.

The Challenge. Developing a flexible control system for a prototype X-ray photolithography source unit, with a very high count of digital and analog channels. The control system should also meet some hardware and software real-time requirements.

The Solution. Using LabVIEW for fast program development, SCXI for analog and digital input/output operations and signal conditioning, plug-in boards for analog signal digitization, control over SCXI modules, interface with a custom X-ray energy digitizer and complicated real-time pulse sequence generation.

Introduction

The Dense Plasma Focus X-ray Source prototype developed by Science Research Laboratory, Inc. of Somerville, Massachusetts, under a contract with a Fortune 100 client (DPF Source) is a photolithography tool for semiconductor manufacturing of the next decade. The tool uses high-voltage high-energy gas discharges to generate X-rays. The tool includes multiple subsystems and has a footprint of approximately $20m^2$. Solidus Integration, Inc. was contracted to develop a control system for the tool. The main requirements for the control system were:

- Control of triggering sequences for X-ray generation
- Data acquisition from X-ray digitizer and energy calculation
- Control and monitoring of the state of multiple subsystems
- Extensive event and datalogging for diagnostics purposes.

Control System

For a control system of DPF Source, we chose an industrial computer running Microsoft Windows NT and the LabVIEW development environment. Our choice of LabVIEW was based on two requirements: a need for a fast development environment and a need to quickly implement modifications reflecting changes in the prototype, which was also under development.

The control system consisted of three top-level loops running in parallel:

- A fast, soft real-time loop for X-ray triggering and energy calculations
- A slow loop for housekeeping—subsystem control and monitoring
- An auxiliary loop for support of interaction with the user.

The fast loop had to support generation of a complicated real-time pulse sequence for triggering gas discharges at a rate of up to 150 discharges per second, data acquisition from an X-ray digitizing unit and X-ray statistics calculations. The discharges were started by the user and had to stop upon delivering a desired X-ray dose.

Each discharge involved generating five pulses with various delays and duration and with a real-time precision (10 microseconds). For this purpose, we chose the PC-TIO-10 plug-in board. It was important for us that multiple counters of this board could be programmed individually, but assigned to the same group and started with one command.

The X-ray digitizer is an instrument developed and built by Science Research Laboratory. The instrument detects signals from two X-ray detectors (PIN diodes) and digitizes the signals into 10 bits each, deriving an analog trigger from one of the signals. The digitized signals are stored in the output registers until the arrival of an external reset pulse. For interaction with this instrument, we chose the AT-DIO-32F board which has enough lines for reading the required 20 bits and supports handshaking. The X-ray digitizer did not provide a request pulse upon completion of digitization, so we had to implement a one-handed handshaking—the request pulse for the DIO board was generated by one of the spare counters of the timing input/output (TIO) board after a discharge, and the acknowledge pulse from the digital input/output (DIO) board was sent to the digitizer in order to reset its sample-and-hold circuitry.

The slow loop of the control system was responsible for controlling all the auxiliary subsystems (charging, cooling, etc.) and monitoring the overall system health. The specific requirements that we had to deal with were a large

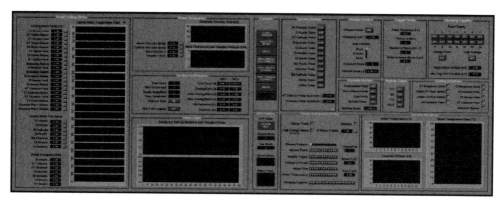

Figure 3-13
User interface of the prototype DPF Source control system

number of analog and digital input and output channels—approximately 200 lines. A combination of the PCI-MIO-16E-1 board and an assortment of SCXI modules were used to address the requirements. The SCXI modules provided high channel counts and optical or galvanic isolation of the front end and the MIO board was used for digitization of analog signals and for controlling the multiplexed SCXI modules.

Finally, the auxiliary loop provided users with the ability to interact with each other through dialog. A separate loop was required here, so that waiting for the user response did not suspend the operation of the rest of the program.

User Interface

While under most circumstances it is advisable not to clutter the user interface with too many controls and indicators, in the case of the DPF Source prototype tool it was important that the development engineers have immediate access to all the data collected by the system. The user interface of the control system spans over two 21-inch monitors (see Figure 3-13).

Controls and indicators were grouped by subsystems, green/red LEDs denoted system status, gray/yellow LEDs showed on/off status of subsystems, and charts showed trends of various analog parameters, like temperature and pressure. While the interface is mostly mouse-driven, emergency shutdown could be initiated by pressing the escape key.

Chapter 3 • Semiconductor Test

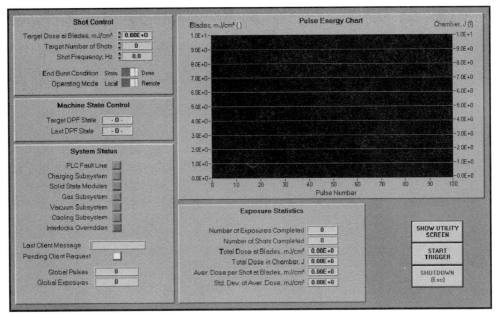

Figure 3-14
Main screen of the prototype DPF Source control system

Conclusion

At this time, the assembly of a production system is in its final stage. The control system of the production model has a few distinctions from that of the prototype. In order to improve system robustness in case of computer or control program failure, we decided to implement control over housekeeping functions using a FieldPoint network and a programmable logic controller (PLC), both of which have watchdog timers and fault detection capabilities and provide isolation of their front ends from the controller. The choice of PLC went to Allen Bradley SLC-500. The PLC communicates with the PC over Ethernet. Support for LabVIEW is provided by using National Instruments Industrial Automation Servers software. For improved noise immunity, analog control and acquisition functions are performed by a network of FieldPoint modules located in proximity to the signal sources and communicating with the PC over the RS-485 interface (at the time of development, Ethernet FieldPoint network modules were not available).

An additional requirement for the production unit is that it should be capable of working both in a stand-alone configuration and as a slave to another instrument (a stepper). The communication between instruments was implemented using TCP/IP over Ethernet.

The user interface of the production tool also had a facelift. The interface is implemented as a hierarchy of screens. The main screen (see Figure 3–14) shows only the X-ray exposure information and subsystem status summary. A password-protected utility screen shows more detailed system information and leads to individual subsystem screens.

Future development plans involve integration of the tool into the client's semiconductor production environment including production database connectivity.

■ Contact Information

Sergey Liberman
Principal Consultant
Solidus Integration, Inc.
26 Wayte Road
Bedford, MA 01730
Tel: (781) 275-5895
Fax: (781) 275-5895
E-mail: s.liberman@ieee.org

Data Acquisition From a Vacuummeter Controlled by RS-232 Standard Using LabVIEW

Iulian Brandea
Electronics Engineer
National Institute for Cryogenics and Isotope Separation
Ramnicu Valcea, Romania

Mihai Culcer
Senior Engineer
National Institute for Cryogenics and Isotope Separation
Ramnicu Valcea, Romania

Dumitru Steflea
Technical Manager
National Institute for Cryogenics and Isotope Separation
Ramnicu Valcea, Romania

Product Used. LabVIEW.

The Challenge. Connecting a microcontroller-based vacuummeter to a personal computer using the RS-232 hardware standard, and the National Instruments LabVIEW and its collection of virtual instruments.

The Solution. Using the software from National Instruments, an instrument driver was created. This provided the customer with a perfect solution for the remote control and data acquisition from an Intel 80CXX microcontroller-based vacuummeter.

Introduction

In our research institute, the Electronics Department designs analysis apparatus, tests and measurements, process monitoring, and control. We make use of digital multimeters, flowmeters, vacuummeters, and mass-spectrome-

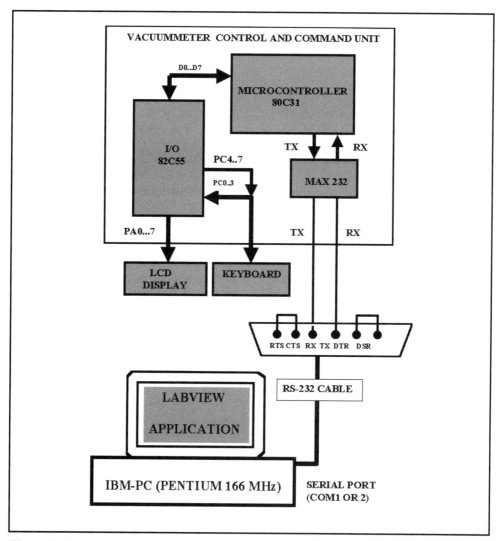

Figure 3-15
Connecting the vacuummeter to the PC via a serial port

ters, manufactured by well-known producers: Hewlett-Packard, Fluke, Keithley, Edwards, and Baltzer. These devices are very expensive and do not meet all of our needs so we had to design and manufacture some of them, according to the main demand: remote control with an IBM-PC. That means the device must be an intelligent one, provided with a microprocessor or a microcontroller. We fulfilled these requirements, building a vacuummeter with a 80C31 microcontroller and two Bayard-Alpert ion gauges, for very low pressures (10^{-3} to 10^{-7} mbar), and low pressures (10 mbar to 10^{-3} mbar). This microcontroller has a built-in circuitry for a serial communication, allowing us to establish a serial communication between the PC (Pentium-166 MHz) and the vacuummeter, according to the RS-232 hardware standard.

Optimum selection of software development tools, however, was not as straightforward. Most producers use the C/C++ language programming tool for developing instrument drivers for their intelligent devices. One of the advantages of C/C++ is its speed, but the compilation and the high-level skill required for optimum programming do not fair well with some requirements, particularly those of versatility, upgradability, and user friendliness. After careful evaluation of several options, we decided to develop a hybrid software package using two different software development tools: LabVIEW and assembly language. We chose LabVIEW because it is dedicated to data acquisition and communications, and contains libraries for data collection, analysis, presentation, and storage. The assembly language for Intel 8051s microcontrollers family is used to write the firmware for the vacuummeter and the communication and arithmetic routines. The hardware connection between the computer and vacuummeter is shown in Figure 3-15.

Front Panel Description

The graphical aspect of the controls and indicators allowed us to build a professional front panel for our vacuummeter. We tried to keep the same look for the VI front panel, like the front panel of the real vacuummeter. The front panel with a virtual keyboard with twelve keys, a simulated LCD display, and two charts is shown in Figure 3-16. This virtual keyboard is used to set the measurement parameters for the vacuummeter, and for starting data acquisition. The main menu is displayed if we click on the key "M". Clicking with the mouse on the keys 1 to 9, the specific submenus are displayed. These submenus consist of:

- Changing the pressure measurement units—mbar, pascal or torr
- Displaying the parameters for ion gauges—electrons current, ions current, and cathode heating current

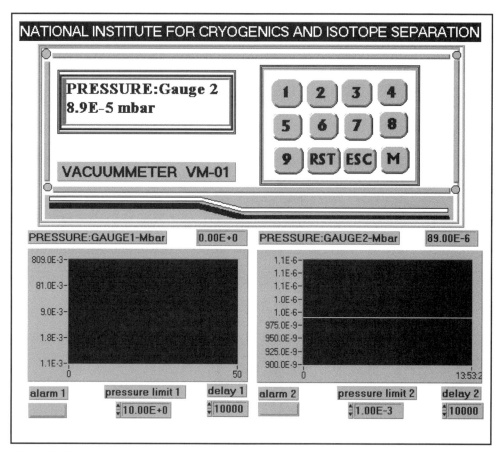

Figure 3-16
The Virtual Instrument front panel

- Selecting the value of cathodic emission current for the ion gauge 2 — 200 µA or 2 mA

- Setting the measurement correction coefficients in respect with the gas (helium, nitrogen, oxygen) at low pressure

- Selecting the working ion gauge — gauge 1, gauge 2, or both

- Ion gauge 2 degassing

- Enabling or disabling the serial communication

- Switching the display backlight on and off

Chapter 3 • Semiconductor Test

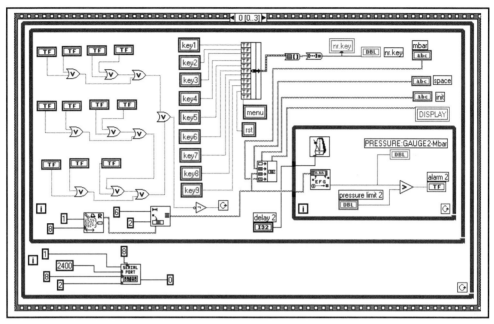

Figure 3-17
Reading the virtual keyboard and receiving data from the vacuummeter

- Displaying status parameters of the vacuummeter — voltage to frequency conversion accuracy, power supplies voltages
- Testing the electrometer amplifier of the vacuummeter.

The front panel is completed with two waveform charts, which display the pressure from the tanks where the ion gauges are placed. Because these charts are used to follow the pressure trend, they work in strip chart mode.

Block Diagram Description

As shown in Figure 3-17, when the VI starts running the serial port (COM2) is initialized at 2400 baud, 8 data bits, 2 stop bits, and no parity. After that, the virtual keyboard is scanned. If no key is pressed, the VI executes while loop in which data are continuously received from vacuummeter. The data are the result of pressure measurements with the ion gauge 2. The received string of data is displayed after the control characters are rejected. For dis-

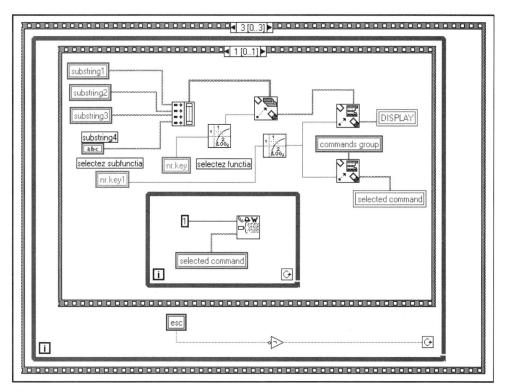

Figure 3-18
Transmitting commands to the vacuummeter

playing the result on the chart, this string is converted into a number (engineering notation).

If any key is pressed except escape, the VI will execute the next frame. Before that, the VI bundles all the Boolean controls and converts the result into a floating number. In frame 2, it extracts the logarithm from this number and obtains a new number which represents the pressed key. In frame 3, shown in Figure 3-18, the virtual keyboard is scanned again to determine which item is selected from the specific submenu. According to this item, a string is written to the COM2 port via the Serial Port Write VI. This string is received by the vacuummeter and a specific task is performed.

Results

The VI that we wrote has many capabilities, including the ability to automatically store measurement results on the hard disk at fixed time intervals. Warning messages are provided for operators when the measured pressure is greater than 10 mbar. In this situation, both ion gauges will be automatically disconnected. The vacuummeter can work as a stand-alone, receiving the commands from the keyboard placed on its front panel. But, when this device is included in a complex system, it is necessary for remote control from the PC. Our LabVIEW software application is very flexible. We can easily change the commands set (coupling the ion gauges, changing the units, setting the limits for higher and lower measured pressure, etc.), which is difficult to do with conventional programming languages.

■ Contact Information

Iulian Brandea
Electronics Engineer
National Institute for Cryogenics and Isotope Separation
Ramnicu Valcea-1000
CP 10, Valcea, Romania
Tel: 040-050/732744
Fax: 040-050/732746
E-mail: brandea_I@yahoo.com

4
TELECOMMUNICATIONS TEST

What is Telecommunications?

Telecommunications can be divided into four major market segments: Telephony, Wireless, Datacom, and Broadcast.

Telephony encompasses the world of the telephone. The traditional Plain Old Telephone System (POTS) connects the analog telephone at your home or business across a pair of copper wires to a telephone company central office. At the central office, an analog POTS signal is digitized and with other signals combined to form T1, a 1.544 Mbps digital service. T1 is then often multiplexed with other T1 signals to a higher-rate T3, or Asynchronous Transfer Mode (ATM).

Manufacturers in the Telephony market need to test their products to verify performance. Complete tests include parametric testing (measuring the characteristics of the signal) and protocol analysis (analyzing the information carried by the signals). Performing these tests can be carried out with digital multimeters, oscilloscopes, spectrum analyzers, and audio analyzers.

The Wireless market is dominated by cellular and personal communications systems, including pagers, two-way radios, and other radio links, which the Federal Communications Commission regulates. The FCC assigns wireless communications on channels assigned to the radio frequency (RF) spectrum, and these RF channels are modulated by lower-frequency baseband signals that can be analog or digital.

In analog wireless, the RF carrier is modulated by a direct analog of the baseband signal. With digital wireless, the baseband signal is digitized and encoded using a variety of techniques such as Time or Code Division Multiplex Access (TDMA or CDMA).

Manufacturers in the Wireless market perform RF signal measurements, baseband signal measurements, and protocol analysis. RF signal analysis is done mainly with spectrum analyzers or specialized communications test sets specific to the wireless service. Baseband testing is done by measuring signal parametrics (like telephony) and by protocol analysis of the information carried by the signals. Protocol analysis may be an integral part of communications test sets (such as cellular or PCS), or it might be performed by discrete protocol analyzers.

The Datacom market consists of local and wide area networks (LANs or WANs). The predominant LAN protocol is Ethernet 10BaseT and 100BaseT (and gigabit is on the way). Nearly every company with more than 20 employees connects computers with a LAN. LAN equipment includes network interface cards, routers, hubs, and switches—all of which play a part in distributing data across a network. LANs connect to other LANs by WANs. Testing in the Datacom market consists of protocol analysis and network performance testing. The majority of this testing is done via RS-232 or Ethernet connections.

Broadband traditionally consists of one-way transmission of radio and television signals directly from transmitters, satellites, or cables.

Similar to wireless, the broadcast market is dominated by RF measurements, which are performed with RF spectrum analyzers or specialized communications test sets. Baseband testing includes parametric testing of the video, data, and audio signals. It also includes protocol analysis, especially with digital broadcast of TV or radio signals.

What are the Present Trends and Challenges?

- **Increasing customers expectations** — With all the new technological advancements, customers demand more features for a lower cost at a higher quality.

- **Blurring lines between segments** — The once clear lines among the four market segments have faded with the emergence of new technologies. For instance, people now can have conversations across the Internet with VOIP (voice over Internet protocol).

- **Higher bandwidth** — New features use more bandwidth than ever before. For example, cell phones that once required limited bandwidth to carry voice signals now have been upgraded to provide Internet access.

- **Increased competition** — More companies are getting on the telecom bandwagon and providing products in several market segments. For example, companies that have traditionally only offered television or cable access now offer Internet access across their cable lines. They directly compete with telephone companies that once held a monopoly on dial-up Internet access.

    The paper entitled "LabVIEW-Based Antenna Measurements" describes how Virginia Tech University Antenna Lab has entered the telecom market with sophisticated, state-of-the-art automated antenna measurement facility equivalent to commercial systems with National Instruments products. And while the lab is not competing for customers, it exemplifies how entering the market is becoming easier.

- **Faster time to market** — Telecommunications consumers do not want to wait for new products. This creates enormous internal pressure to get new products out the door faster. Products that used to be developed in one year are now needed in half that time.

    The paper "Telecommunications Protocol Analysis Tool" describes how a senior engineer at Alliance Technologies Group, Inc. helps shorten development time with a LabWindows/CVI application with graphical and textual output that executes on both Sun and PC computers.

What is in the Future?

Consumers can look forward to seeing a variety of affordable devices that combine a variety of features found on today's popular telecommunications products. One, for instance, is wireless personal data assistants, which will combine a palm-size computer, cell phone, data port, and even a television on one low-cost handheld device.

Products like PDAs will be feature-rich, and reconfigurable on the fly. Testing for these types of devices will need to be more flexible and modular. National Instruments approach to PC-based hardware and software lends itself well to the test and measurement demand of the future.

How is National Instruments Involved?

While it may not be possible to control the trends and challenges facing the telecommunications industry, it is possible to choose tools that companies can use to develop, automate, and execute tests faster. National Instruments addresses these concerns and currently provides one of the most popular software development environments used in telecommunications test. National Instruments LabVIEW is used to create test and measurement solutions to major telecommunications companies in all three large telecommunication regions in the world: the U.S., Europe, and Asia. Other National Instruments hardware and software products also can be combined to create computer-based measurement and automation solutions for the telecommunications industry.

LabVIEW-Based Antenna Measurements

Kevin Mescher
Satellite Communications and Antenna Group
Virginia Tech University

Product Used. LabVIEW.

The Challenge. Creating a sophisticated, state-of-the-art automated antenna measurement facility.

The Solution. Using LabVIEW to provide a convenient way to control hardware, process the data, and provide a user-friendly interface for the program.

Introduction

As a result of the explosive growth in wireless communications, the design and testing of antennas has taken on a renewed importance. One of the important performance characteristics of antennas is its radiation pattern. These plots describe how the power radiated varies with direction around the antenna. Pattern plots are used to assess the performance of the antenna.

Traditional antenna measurement systems employ analog instruments and analog data collection methods. In fact, many systems still in use today have pen and ink recorders for recording radiation patterns. Movement toward digital systems is slowed by the cost of new instrumentation. As a compromise, many systems have replaced the analog data output with digital output, usually using a personal computer for data acquisition and data processing. Some systems have relatively sophisticated data manipulation and display modules in the PC.

Virginia Tech University uses an antenna range to test its antennas. A typical range consists of a signal generator, source antenna, and a receiver. The source antenna transmits a test signal which is received by the antenna under test (AUT) located some distance away. The AUT is rotated and the received signal is recorded to produce a radiation pattern plot of signal strength versus azimuth angle. All hardware must be controlled at the same time, and much of the hardware must be reset for the next measurement. An output device is required to plot the radiation patterns of the antenna. This device can be a plotter connected directly to the output of the receiver, but the preferred approach is to connect a printer to a PC, permitting the operator to command and control the printouts.

Multipath Distortion

A major difficulty encountered when trying to measure antenna patterns is a phenomenon called multipath distortion, in which unwanted reflections of the transmitted signal arrive at the AUT and interfere with the direct signal. The multipath signal distorts the measured antenna pattern. However, multipath effects can be reduced by using a large, open outdoor test site for the antenna range or by making measurements inside an anechoic chamber. An anechoic chamber has walls that absorb RF radiation, reducing the reflected signals. An outdoor test site usually requires a large open area. Many outdoor antenna ranges cover several acres or more. Anechoic chambers are specially built rooms, usually metal shielded, with RF-absorbing blocks of foam covering all the interior surfaces. Both outdoor ranges and anechoic chambers can be expensive to build and operate.

Virginia Tech University's antenna range is on the roof of a campus engineering building, which provides a secure open space but also has objects that can create multipath distortion. The antenna range has digital data collection and display control using LabVIEW. Using LabVIEW, signal processing techniques can be applied to the pattern measurement data to reduce the range multipath distortion effects. In order to mitigate the multipath effects, we use time domain processing of the received signals.

Hardware Control Software

In 1995, a LabVIEW-based program called Range Runner was written to handle all the hardware control for Virginia Tech University's outdoor antenna range. It measures the antenna radiation pattern and provides pattern printouts via a laser printer. It also has the ability to perform automated time-domain processing. Andrew Predoehl, a graduate student in the Antenna Group, wrote the first version of Range Runner.

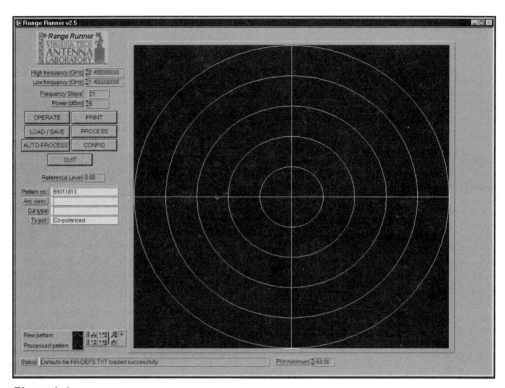

Figure 4-1
Range Runner main screen

A characteristic of multipath distortion is that all the reflected signals arrive at the test antenna delayed in time from the direct signal. By using time domain processing, these can be filtered out of the measured pattern. Range Runner automates this process and gives the user full control over the processing.

One of the limitations of the original measurement system is that the actual measurement process was very slow due to the method for controlling the equipment. The range uses a Scientific Atlanta 1783 receiver and HP8648C and HP8673B signal generators. The signal generators and the receiver are separate components, and each must be controlled individually. Collecting about 3,780 data points took about 2½ hours. After data was collected, the processing step was also slow, due to the large amount of data to process.

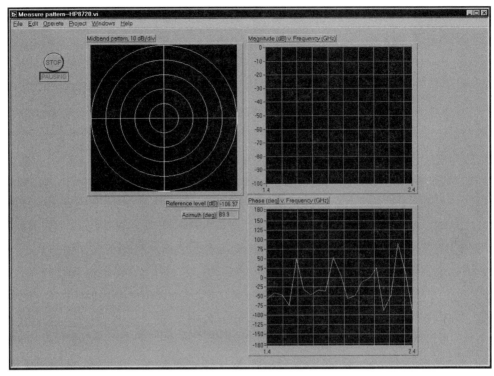

Figure 4–2
Range Runner measurement screen

Results

The Virginia Tech University Antenna Laboratory is completing a new anechoic chamber and will be using modern digital instrumentation based on a HP8720C vector network analyzer. In anticipation of the new chamber, Range Runner was updated to operate with the network analyzer. The new version of the software is much faster, and the new system was able to collect 36,180 data points in about five minutes, a speed increase of about 30 times. The new software also allows calibration of the network analyzer and allows for more types of measurements.

Using LabVIEW the Virginia Tech University Antenna Lab has created a sophisticated, state-of-the-art automated antenna measurement facility. This system is equivalent to commercial software, which can cost tens of thousands of dollars. LabVIEW made all this possible by providing a convenient way to control hardware, and process the data, and provide a user-friendly interface for the program.

References

[1] Andrew M. Predoehl and Warren L. Stutzman, "Implementation and Results of a Time-Domain Gating System for a Far-Field Range", *Proceedings of Antenna Measurement Techniques Association (AMTA)*, (Boston, MA), pp. 8-12, November 17, 1997.

■ Contact Information

Kevin Mescher

>Satellite Communications and Antenna Group
>Virginia Tech University
>347A Whittemore Hall
>Blacksburg, VA 24060
>Tel: (540) 231-6834
>Fax: (540) 231-3355
>E-mail: Kmescher@vt.edu

Dr. W. L. Stutzman

>Director, Antenna Group
>Virginia Tech University
>627 Whittemore Hall
>Blacksburg, VA 24061
>Tel: (540) 231-6834

Telecommunications Protocol Analysis Tool

Michael Tanquary
Senior Engineer
Alliance Technologies Group, Inc.

Product Used. LabWindows/CVI.

The Challenge. Developing a protocol analysis tool that executes on both Sun and PC computers for the software development of the Motorola Satellite Series 9500 portable telephone used for the Iridium system.

The Solution. Using the multiple platform capabilities of LabWindows/CVI to develop an application with graphical summaries and textual output.

Introduction

Digital cellular systems utilize a layer-based protocol to transfer information between the telephone and the network to which the telephone connects. Each layer implements a functional component of the protocol. For example, the second layer helps ensure the reliable transfer of data between a Satellite Series phone for the Iridium system and a satellite. The implementation of the protocol is handled by real-time software that executes on a microprocessor in the telephone. During development of this software emulators and source debuggers are used to test the software. Some tests, though, require the software to operate without any intrusion to help ensure that millisecond response times are maintained. In these situations the software developer relies on logged messages generated by the software during execution for debugging.

The messages are sent from the phone to a host computer in binary form to maximize the amount of data that is logged. The data log contains both network-related and internal, product-specific information useful in debugging the phone's software. The software developer analyzes the data log after the test is complete to ascertain what events occurred during the test.

Prior to the LabWindows/CVI-based solution, the software developer would examine a textual version of the data log. Typically, a person is only interested in a subset of the messages dealing with a particular layer among all the other messages present. By using LabWindows/CVI the same textual output capability was maintained, but several new easy to use graphical representations were added. These new graphs allow users to focus on their area of development and present the data in a logical manner.

System Description

To help ensure platform independence, we developed a method of converting the data in the binary log file to a representation that mirrored the C structure that defines the format of the message. The software that performed this conversion operated on the Sun and PC computers without modification or conditional compiling. By formatting the data into the C structure representation, we were able to use the structure fields to easily process and analyze the data log.

Given that we could now process the contents of all messages in the data log, we set out to create user-interfaces and analysis routines to present the data from the log in a logical manner. We chose to present data related to each layer of the protocol stack. The Iridium protocol stack has three layers that software developers need to analyze:

- **Layer One (Physical Layer)** — includes unacknowledged communication and data used to establish an initial link to the system
- **Layer Two (Data Link Layer)** — messages used to manage the transfer of higher layer messages to help guarantee they are received by the receiving entity
- **Layer Three** — messages used to manage the communication link, maintain the phone call, and many other tasks.

For layer one analysis the software developers need to see a bounce diagram showing the messages sent between the phone and the satellites, communication trail showing the handoffs the phone made during the phone call, and the link statistics of the phone call. An example of the communication trail is presented in Figure 4-3.

Chapter 4 • Telecommunications Test

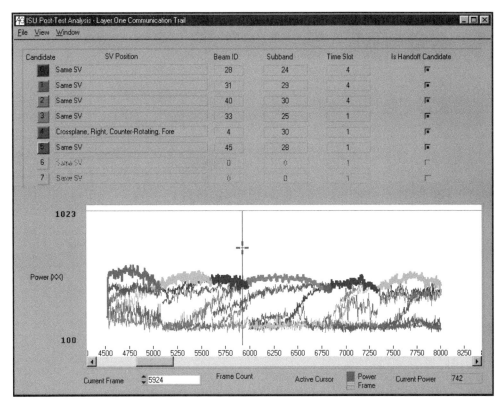

Figure 4-3
Layer one communication trail

For the Iridium system, the movement of the satellites is the main cause of handoffs, not the movement of the mobile telephone. Because the satellites are moving, the cells, called beams in the Iridium system, move across the surface of the earth. The telephone has to handoff to a new beam once the current beam's signal weakens due to its movement away from the caller. A software developer would like to view the handoff events to see if the phone operates properly (i.e., did the phone handoff to the best beam).

Figure 4-3 represents the satellite's beams as different colors on the graph. The x-axis represents time and the y-axis represents the power of the beam. The upper section of the screen in Figure 4-3 lists the possible handoff candidate beams for the phone.

By looking at the graph in Figure 4-3, a software developer could see that the telephone performed handoffs so it could remain on the most powerful beam, thus obtaining the best call quality available.

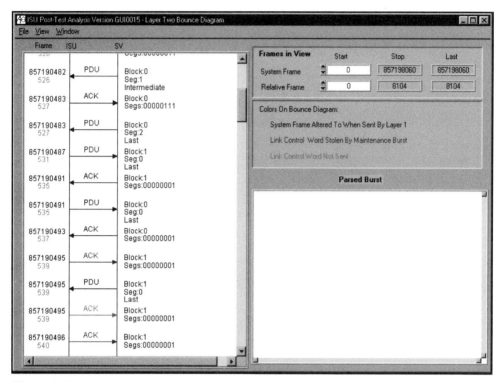

Figure 4-4
Layer two bounce diagram

For layer two and layer three analysis, the user needs to see a bounce diagram that shows the flow of messages between two peer entities. Figure 4-4 presents a layer two analysis screen that contains a bounce diagram showing the messages that were transferred during a test. The arrows represent information going from one entity to the other, with the time the message was transferred to the left of the arrow. To the right of the ladder is information that identifies the message. Further details of the contents of the message can be viewed in the Parsed Burst text box by selecting the arrow (contents are not shown since information is confidential).

Results

LabWindows/CVI was crucial in providing a graphical analysis application that can operate on a Sun and PC platform using the same source code. The flexibility of LabWindows/CVI allowed Motorola to create the application from a single version of software stored in a location accessible by both Sun and PC computers. Motorola developers now have an effective tool to analyze the operation of their phone in the lab environment or out in the field.

■ Contact Information

Michael Tanquary
Senior Engineer
Alliance Technologies Group, Inc.
1335 Wilhelm Road, Suite B
Mundelein, IL 60060-4488
Tel: (847) 247-9284
Fax: (847) 247-9724
E-mail: mtanquary@atgroupinc.com

Chapter 4 • Telecommunications Test

Remote Diagnostics in a Fiber Optic Network

Russell Davoli
Project Quality Manager
G Systems

Product Used. LabVIEW 5.0.

The Challenge. Controlling a fiber optic wavelength division multiplexing (WDM) network tester that is installed in a field monitoring facility for a fiber optic network from a centralized facility. Newer versions of the tester do not have a front panel, so the remote control program is the only way to control the instrument.

The Solution. Implementing a LabVIEW virtual front panel for the WDM network tester and controlling the tester via the GPIB or serial interface. To control testers installed at very distant locations, the application can connect through modems and communicate over the serial interface.

Introduction

The Anritsu Company serves telecommunications customers who have large networks with numerous far-flung monitoring substations. In these substations, the company installs racks of diagnostic equipment to monitor the health of the network. When a problem arises, a trained technician troubleshoots the problems with the same diagnostic equipment. The technician must travel to the monitoring substation, which can be time-consuming and expensive. To save time, Anritsu wanted to provide its customers with a remote control capability for its instruments in the diagnostic rack.

The primary instrument in such a rack is the Anritsu MS9720A, a wavelength division multiplexing (WDM) tester. The MS9720A is a valuable tool because it integrates an optical spectrum analyzer with specialized WDM analysis functions. The instrument implements nine different measurement modes, including the spectrum analyzer, WDM tester, and other measurements of optical transmission quality. Each mode displays its own set of graphs, tables, menus, and other display features on the built-in cathode ray tube (CRT) on the front panel. The remote control program would have to duplicate the front panel interface as closely as possible to allow technicians already familiar with the instrument to use the program with minimal retraining.

Anritsu was also looking for ways to meet customer desires for lower-cost equipment solutions for the diagnostic racks. People don't typically spend much time in front of the instruments, so Anritsu is producing versions of the MS9720A that do not have the front panel CRT, knob, and buttons. Because the remote control program already duplicates the front panel interface, the program would serve as the virtual front panel to the new instruments.

Program Design

The remote control program had to implement front panels that are very similar to those of the actual instrument. The front panel of the program is shown in Figure 4-5. Making such virtual front panels is straightforward in LabVIEW, so this was the natural choice for the development system. We chose to implement the application in three layers: the user interface, the instrument proxy, and the instrument driver.

At the top level, the user interface implements the front panel controls and indicators and processes the user input from the controls and menus. Because the virtual front panel is running on a standard PC running Microsoft Windows NT, 98, or 95, we took advantage of the keyboard and mouse to simplify some aspects of the instrument front panel. For instance, the instrument uses the knob in many cases to aid in entering numbers or in selecting letters to form a title. On the computer, we eliminated the knob and allowed the user to type numbers into numeric controls or enter text into string controls. As a beneficial side effect, this freed up more screen space for displaying data. Another user interface improvement came from the ability to drag cursors on the graphs with the mouse. This LabVIEW graph feature simplified the programming, as well as smoothed the user's interaction with the cursors because he can simply click and drag the cursor rather than go through the tedious selection process required by the more limited interface on the instrument.

The user interface code calls the instrument proxy layer to control the instrument and get its settings and data. The purpose of the instrument proxy is to isolate the user interface from the details of sequencing the com-

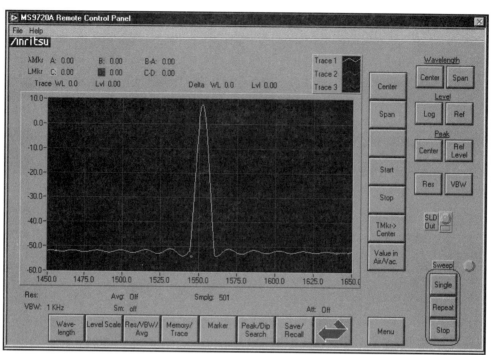

Figure 4-5
The virtual front panel for the remote control program

mands needed to control a particular mode. Each measurement mode also keeps a nonvolatile backup of its last settings, so when the user changes modes, the new mode sets the instrument to those settings. The proxy layer is responsible for retrieving all of the settings of the current mode, which happens after changing to a new measurement mode or connecting to the instrument when the program initializes.

The instrument driver layer is the familiar concept of a set of virtual instruments (VIs) that send commands and queries to the instrument. We implemented the instrument driver using a driver software architecture called VISA as National Instruments recommends. This gave us an easy way to meet the key remote communication design goal by taking advantage of the GPIB and serial communication capability built into VISA. For use and testing with the instrument beside the computer, GPIB is very convenient. When remotely controlling the embedded instruments, the technician calls from his desktop computer to a modem attached to the instrument, so the program must use serial communication. The instrument driver VIs did not need to change to accommodate both communication modes. To implement the modem control, we adapted VIs from an article in *LabVIEW Technical Resource* Volume 5, Number 1, entitled, "Attention: Paging All LabVIEW Users."

Results

LabVIEW and VISA gave us a powerful foundation on which to build our application. The VISA functions allowed us to build the drivers and the application on top using the GPIB communication mode, then switch over to serial mode very late in the project with only the minimal work of changing the communication initialization function. The front panel controls and indicators, along with the power of attribute node programming, gave us the right set of tools to implement the virtual front panel. The success of this approach was proven when we gave the program to our first beta tester, who was immediately able to use familiar functions of the instrument.

■ Contact Information

Russell Davoli

> Project Quality Manager
> G Systems
> 6060 N Central Expwy, Suite 502
> Dallas, TX 75206
> Tel: (214) 373-9494
> Fax: (214) 706-0506
> E-mail: russell_davoli@gsystems.com

Quick Real-Time Test of Communication Algorithm Using LabVIEW

Fred Berkowitz
Advanced Micro Devices

Whu-ming Young
Advanced Micro Devices

Products Used. LabVIEW, GPIB.

The Challenge. Implementing and optimizing a new system algorithm at the earliest stage possible before any expensive hardware implementation. Modern communication systems rely on complex, computation intensive signal processing algorithms to achieve high performance under noisy environments. The system algorithm simulation results are limited in scope and cannot realistically cover all the major cases.

The Solution. Using LabVIEW software together with GPIB as the heart of a system emulator to integrate high performance instruments. LabVIEW can take the complicated sequence generated from math tools (e.g., Matlab) and transmit it to the real channels via an arbitrary waveform generator. Likewise, LabVIEW can take the received signal into the PC for further algorithm processing/analysis. The result was a reliable, inexpensive tool that provides performance evaluation and optimization of new algorithms on the real medium and environment, directly from the software model, without recurring to expensive and time-consuming hardware implementation.

System Hardware Control

LabVIEW and a GPIB card are installed in a 350 MHz AMD K6 PC to control the whole hybrid digital communication system. A waveform generator and a digitizer are used to perform transmitter and receiver functions, respectively. The signal stream is generated in the PC from our proprietary simulation models and downloaded as a data file. LabVIEW can direct the GPIB card to transfer the data file to the memory of the waveform generator. Furthermore, LabVIEW also sets up the 784 as a receiver to collect the data transmitted by the generator. The resulting acquired signal is finally uploaded for further processing in an ASCII format file.

A communication system requires a wide range of control parameters to enable its full operations. LabVIEW plays a key role in this regard. For 12-bit D/A and A/D rates with 50 Msample/s range, LabVIEW is used to transfer large data files and coordinate a large number of control parameters. Additional functions achieved by LabVIEW are timing and logical control of the test loop since it has to cooperate with an external application that generates the input stream and accepts the output samples, and has to do it in an efficient way. The loop has to repeat enough times to provide a statistically meaningful sample.

Benchmark of System Costs

Traditional communication development systems consist of transmitter and receiver hardware, control software, and signal processing hardware/software. Since it is on the high end of the performance curve, all these pieces are custom built with very little re-usability from one design to another. Typically, at least 2-3 hardware renovations are needed to increase either the receiver sensitivity or total system robustness even within the same generation of the digital communication system. Consequently, only the military or large corporations can afford this type of research and development projects.

With the rapid advancement of PC and PC-based virtual instruments, such as National Instruments, the cost and system development cycles for digital communication start to change. We did a benchmark in terms of system cost and development time.

System Cost Reduction

Traditional communication system development is "test-board" intensive. (After the system algorithm and architecture are finalized, the final production board is still needed in both traditional and out system, which is outside of the scope of this paper.) Assuming, on average, each system takes four board revisions to finalize its architecture; each board on average costs around $10,000. The total system development cost from hardware boards is around $40,000 and is not reusable. On the other hand, the hardware in this system is an arbitrary waveform generator ($5,000), a digitizer ($7,000), a computer loaded with LabVIEW, and a GPIB card ($3,500). The total hardware cost is $15,500. Not only is it around 40 percent less than the original price, but also it is re-usable. So far, we have used this system to develop two totally different communication systems; therefore, the saving is more than 80 percent (or the cost is only 20 percent) of the traditional way ($80,000 for two systems).

Development Time Reduction

The time saved on development comes in two areas: 1) reduction of board turn-around time and 2) reduction of control software development time. Because the new development system does not require a new board each time the systems performance needs tuning, the test board turnaround time is totally eliminated. On average, the turnaround time of a test board is two weeks, so a total of eight weeks is saved using the new LabVIEW PC-based communication development system.

Another gain in the reduction of development time is in the control software, since the development platform is more stable than the one used in traditional methods. Normally, it took about four weeks to develop the control software for each generation of the communication system. Now, it takes about one week, with most of the time spent in the area of updating the whole system control in LabVIEW. The improvement time is around 66 percent.

Results

A PC-based development system for digital communication has been produced. It has been used successfully in developing two different digital communication systems. Saving on the cost of the system hardware coupled with the reduction in the development time further enhances the value of computer-based measurement and automation. It is used not only for testing established products, but also in developing new algorithms and architectures. This paper confirms the feasibility and value of LabVIEW-based systems in replacing traditional, expensive communication system development methodology. In our two projects, our management team is pleased to see the huge savings in the cost of the system (approximately $64,000 or 80 percent less) and reducing the software development time (approximately 6 weeks or 66 percent less) from using the computer-based transmitter and receiver systems. This application paper demonstrates an excellent example for using virtual instrumentation in developing digital communication system.

■ Contact Information

Fred Berkowitz
Advanced Micro Devices
Tel: (408) 749-3042
Fax: (408) 749-5439
E-mail: fred.berkowitz@amd.com

Whu-ming Young
Advanced Micro Devices
Tel: (408) 749-3042
Fax: (408) 749-5439
E-mail: William.young@amd.com

Common Test Software for Cellular Base Stations

Jim Morrison
Lead Engineer
Network Solutions Sector
Motorola, Inc.

Chuck Patterson
Senior Software Engineer
Network Solutions Sector
Motorola, Inc.

Products Used. TestStand™, LabVIEW, LabVIEW SQL Toolkit, PCI-GPIB Cards.

The Challenge. Developing a comprehensive test software application capable of testing a wide variety of Motorola cellular base station products (both current and future), each with differing test procedures and specifications. The application must be modular and adaptable, and accommodate differences in instrumentation models and brands across test benches.

The Solution. Creating a database for test specifications and test procedures in ORACLE, then developing a database-driven test application with interchangeable instrument and product driver technology using National Instruments TestStand, LabVIEW, and the LabVIEW SQL Toolkit.

Introduction

Motorola's Fort Worth Integration Test and Certification (ITC) and Assembly Test Engineering (ATE) test software groups teamed up to develop a common solution with the intention of bolstering software innovation and maximizing collective reuse amongst the two groups. Primarily using National Instruments TestStand and LabVIEW, ITC and ATE together effectively developed a database-driven, standardized test executive with common software classes. The test application adapts to differences in products and test environments.

Requirements

Motorola recently embarked upon a software process improvement effort within test software groups. This effort, coupled with a desire for standardization and conformity, resulted in the idea of common test software for all Motorola cellular base station products. The nucleus of the ITC and ATE teams' endeavors was common; each group tested the same products with similar tests. The main challenge was to bring together the specific differences between products and test demands of development and manufacturing. The goals of the project were:

- Create a common set of software that could be used by both ITC and ATE
- Standardize software process within and across the ITC and ATE groups
- Maximize reuse while minimizing duplicated effort
- Create effective training and documentation
- Reduce manpower per product test capability
- Reduce future time-to-test
- Enable a technology roadmap for test applications.

Representatives from each group began reviewing the existing ITC and ATE test code with these established goals in mind. After reviewing several of these product test applications, the following conclusions were drawn:

- A common test executive was needed
- Product and test-station personalities must be kept external to the test software
- Looping structures must be adaptable and interchangeable
- Different brands or models of similar instrumentation must be interchangeable without software modification

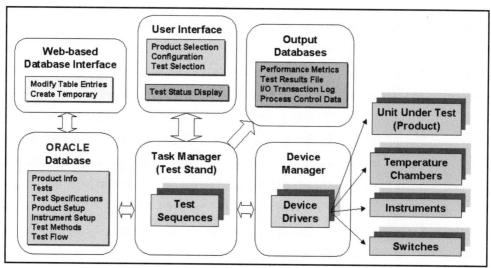

Figure 4-6
Motorola's Common Test Software Model

- State caching must be implemented to ensure faster and more efficient instrument and product communication.

Software Model

These requirements led to the development of a Common Test Software model (Figure 4-6). Key components include a centralized test executive, a smart task manager for looping, interchangeable device and product drivers, and an external database for product information and test specifications.

The team chose National Instruments TestStand as the test executive. TestStand is an ActiveX-based test executive with prototype adapters for LabVIEW, LabWindows/CVI, Visual Basic and C++, and DLLs. Software modules from any of these languages can be combined or used exclusively to create the test routines. The executive itself provides many necessary services such as login with definable permission levels, report generation, a development environment, process models, and database management. TestStand was chosen for its maximum flexibility.

The ITC and ATE test groups were already proficient LabVIEW programmers, so National Instruments LabVIEW graphical programming language was a natural choice for the test language. This language is tailored to test automation and contains many built-in functions ideal for developing tests. In addition to all of the normal primitives of any software language, LabVIEW offers built in multi-tasking, multi-threading, GPIB communication, serial communications, statistical processing, notifications, semaphores, rendezvous, and queues. LabVIEW is a graphical programming language that is simple to code, offering maximum productivity.

The team chose an interchangeable state-caching driver approach based on the concept of National Instruments Interchangeable Virtual Instruments (IVI™). However, IVI is a new product and is not readily available for the specialized instrumentation required for cellular base station testing. The group developed drivers such that they could be replaced with IVI once those drivers become available.

Instrument and product interchangeability was accomplished by defining instruments and products as classes of objects. The classes define a certain instrument's functions from a perspective that is applicable to all of the models and brands in its class. Within each instrument class, function drivers are designed with the defined inputs necessary of that function. Beneath the class layer, model and brand specific function drivers are used to convert the class function to the actual instrument function. These classes can then point to any instrument defined without changing the test software. This modular approach allows adding new brands and models with minimal impact to the core application.

In a majority of test applications, the largest bottleneck is the communications between the test station controller and the product under test and instrumentation. Reducing communication calls between the test station controller, the product, and the instruments can dramatically reduce test time. State caching monitors current configurations of the product and instrumentation so that redundant setup calls are not sent across the bus.

Motorola recently chose ORACLE as a corporate standard, so the team created a relational database in ORACLE. A database map (Figure 4-7) was used to relate tables of product parameters and specifications to test procedure tables and test bench configuration tables. Using a relational database not only centralized all of the pertinent data, it also reduced the amount of redundant information. The team developed a web-based ORACLE interface such that operators and developers alike could access the data from any workstation on the Windows NT domain.

The application is capable of running with or without a connection to the ORACLE database. Prior to each product test run, the application copies all

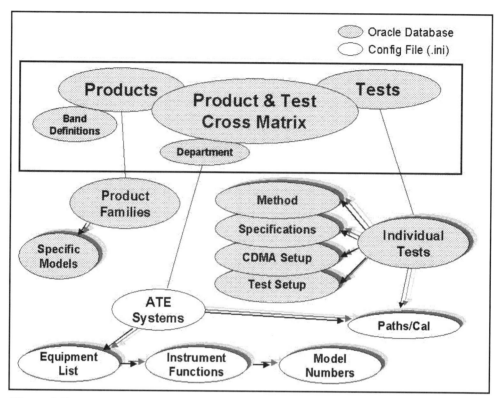

Figure 4-7
ORACLE relational database map

necessary test tables for the specified product type to local files in the configuration file format (.ini). As long as the same product type is being tested, the application can continue testing even if the ORACLE database is down. The application may be set to run in a local only mode once the database files have been loaded. In integration testing, it is often very useful to use the local copy of the test files as a scratch pad for modifying specification tolerances or configuration parameters. Permission levels determine which operators can switch to local mode, and all local files are permission protected.

When performing certification and manufacturing functional verification on cellular base stations, many tests are repeated while varying key parameter values. Examples of these dynamic parameters include supply voltage, receive frequency, transmit output power, channel, and sector. The task manager for this application reads a test register that contains index fields for each of the necessary looping parameters. The indexes correspond to a value

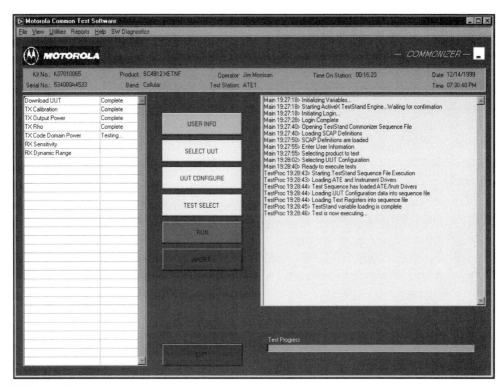

Figure 4-8
Main user interface

for the associated parameter. In order to permit flexible loop control over these parameters, test registers are grouped into a test profile. A test profile editor permits easy creation of multiple test registers. For each profile, execution of the test registers may be specified explicitly or a register sort may be selected in order to optimize test time.

The main user interface (Figure 4-8) permits the test operator to enter personal data, select product type and configuration, select tests to perform, and monitor test progress. Once a product is selected, the database is queried to determine the appropriate configuration and test selection options.

Using TestStand, LabVIEW, the LabVIEW SQL Toolkit, and ORACLE made accomplishing each of the above essential pieces of the common test application possible. With a common test executive and complete product and instrument drivers, creating new tests becomes a simple matter of modifying a standard test template. In addition, the modular design of this test application permits the addition of new tests, new instruments, and new products in a relatively short amount of time.

Estimated Return on Investment

Creating a common test application equates to a large reduction in cost. The ITC and ATE groups currently maintain a total of eight separate test applications. By combining efforts between the groups and developing a single modular test application, the combined annual maintenance cost for the groups can be reduced from $700,000 to $400,000 and the annual projected new product test development cost from roughly $200,000 to $25,000. The maintenance cost reduction is based on the number of engineers needed to support the multiple applications versus the number required to maintain a common application over a single year. Development cost is based on previous test application effort expended at a rate of one new base station product family per year. The development and maintenance cost reductions combine for a total reduction of $475,000 per year.

Results

Using commercial, off-the-shelf software permitted us to concentrate on developing tests rather than a test executive. The primary benefits of developing this common test software solution are:

- Reduced cost of future development and maintenance
- Eliminated duplication of efforts and maximized reuse
- Centralized all specifications and procedures into a single database
- Provided a path for implementation of coding standards and process improvement
- Permitted common training for test developers
- Placed all test software development under configuration management
- Reduced time-to-test for product variations and new tests

■ Contact Information

Jim Morrison

        Lead Engineer
        Motorola, Inc.
        5555 North Beach, M/S 7C
        Fort Worth, TX 76137
        Tel: (817) 245-7076
        Fax: (817) 245-6851
        E-mail: QJM003@email.mot.com

Chuck Patterson

        Senior Software Engineer
        Motorola, Inc.
        5555 North Beach, M/S 7C
        Fort Worth, TX 76137
        Tel: (817) 245-7164
        Fax: (817) 245-6851
        E-mail: QCP002@email.mot.com

GENERAL TEST

Why is General Test Important?

The general test and measurement market is worth more than $7 billion in annual sales and grows at about 7 percent annually, according to market research by Prime Data.

Stand-alone test instruments began interfacing with computers in the mid-1960s. These systems created a need for non-standard hardware and software interfaces. IEEE-488 became the standard interface bus in the early 1970s. The IEEE-488 interface bus standardized the hardware interface between test instruments and computers. Around the same time, digital circuit testing created a need for automatic test equipment for digital test. With the introduction of the microprocessor, test instrument manufacturers began including more intelligence in their products.

Today, complex devices create the need for integrated test capabilities. With integrated test, manufacturers build a device or system that is completely testable, resulting in less complex—and less expensive—test equipment. Due to fundamental changes in the market driven by commercial applications, the test instrument market has entered a new era. Devices need to be tested under varying conditions with a variety of signals. It is common to test a unit with analog, digital, video, and protocol signals all in the same tester.

What are the Present Trends and Challenges?

The overall test instrument market is affected by trends that impact instrumentation solutions. The papers included in this chapter reflect the trends listed below.

- **Growing Role of the Personal Computer in Test** — In the last few years, the PC shifted from a peripheral role to a central position in test. Application software and operating systems are gaining in performance. Microsoft Windows OS now contains many features including security, multitasking, and multithreading previously found on more expensive workstation and mainframe computers. Specialized workstations and expensive mainframe machines previously handled the complex applications that simple desktop PCs now cover. Low-cost PCs will continue to drive the performance of test applications.

    The paper entitled "Controlling Aeronautical Hydraulic Actuator Testing with LabVIEW" demonstrates the move of the PC to the center point of the application.

- **Increasing Use of the Internet** — Measurement technology is increasingly integrated with Internet technology. Sharing of data and test programs becomes easier with the Internet. Test results can now be posted to the Web, as they are available, for other users to view. Remote control and distributed execution are possible at lower costs. The common communication platform of the Internet eliminates the cost of supporting a network.

    Gathering data via a network, or remote measurements, can be done with a single machine connected directly to the device. But sometimes measurements must be taken at a remote site. With Ethernet-based measurement devices, users overcome distance limitations.

The Internet also facilitates distributed execution by improving the performance of a system, increasing flexibility, and connecting computers in a network to create a coordinated measurement system. Users will rely on distributed execution to perform various tasks across several different computers.

"Automating the San Francisco Bay Model with LabVIEW" shows integration of motors for controlling tidal flows with remote data acquisition, tied together with the Internet. Also, the papers "LabVIEW-based Interactive/Visual Machines and Drives Teaching Laboratory" and "LabVIEW-Based Interactive Teaching Laboratory" integrate test and measurement with Internet technology to create a distance learning solution.

- **Decreasing Test Cost** — In the late 1980s and early 1990s, the declining growth of the test instrument market resulted in increased competition among instrument makers. According to a 1997 Prime Data Industry Study, "The decline in test measurement prices is a result of new test instrument products that provide increased performance at lower prices and of price discounts on existing products." As the primary platform for test and measurement, the personal computer, drops in price, the cost of test also decreases.

- **Increasing Integration of Test Instruments** — All areas of electronics are now integrating complex interactions between hardware and software. Combining many technologies within one system requires an integrated test solution. This has changed individual test instruments and tools and driven the demand for multifunction test instruments. Multifunction test instruments will make for a better solution because they increase productivity and can solve test problems that cannot be solved with individual test instruments.

As products and solutions become more complex, they require a better range of signals and devices. Each application in the following chapter requires a number of signals and devices to come together.

"LabVIEW Tests M1A1 Ammunition" describes an application combining imaging, wind speed and direction, barrel pressure, round velocity, and other measurements.

- **Changing Technology Requirements** — Historically, fast-changing technology requirements were the biggest factor in the purchase of new test instruments. Applications-oriented technology changes are replacing the technology advances of the past. Products to be tested are growing increasingly complex, requiring greater integration of the test equipment for analog, digital, and software subsystems. Manufacturers must provide solutions for integrated systems because electronic systems cannot be tested properly by testing individual subsystems alone.

  "LabVIEW-based Automation of a Direct-Write Laser Bean Micromachining System" shows the integration of images, motors, and lasers into a single application. "LabVIEW-based Automation of a Direct-Write Laser Beam Micromachining System" shows integration of images, motors, and lasers into a single application.

- **Increasing Numbers of Communications Applications** — The communications industry has replaced military/aerospace as the largest user of test instruments and is driving the future test industry growth and direction. Because communication applications are becoming increasingly complex and combine many technologies for faster communication transfer, there is a demand for multifunction test instruments that test the interaction of the functions in complex systems. The military/aerospace segment is shifting to commercial off-the-shelf solutions (Prime Data Study). See the "Telecommunications" section (Chapter 4) for communications applications.

What are the Future Trends and Challenges?

The worldwide test instrument market will approach $11 billion in 2001, according to Prime Data. Products to be tested will become more and more complex due to a convergence of telecommunication, computer, and other technologies. The test market must provide application-specific/solution-oriented systems for these new markets.

How does National Instruments Fit In?

Instruments have always taken advantage of widely used technologies. In the 19th century, the jeweled movement of the clock was used to build analog meters. In the 1930s, the variable capacitor, the variable resistor, and the vacuum tube from radios were used to build the first electronic instruments. Display technology from television contributed to modern oscilloscopes and analyzers today. Personal computers contribute to high-performance computation and display capabilities at an improving performance-to-price ratio. Beyond the personal computer, test platforms are adopting other technologies from the Internet and cellular phones.

To address the challenges in general test applications, National Instruments offers virtual instrument tools, both software and hardware, for increasing the integration of test systems. Lowering the cost of the test and raising the productivity of the test engineer are additional benefits gained through the use of PC-based virtual instrumentation.

LabVIEW-Based Interactive Teaching Laboratory

Nesimi Ertugrul
Department of Electrical and Electronic Engineering
University of Adelaide
Adelaide, Australia

Products Used. LabVIEW 5.1, HiQ, and AT-MIO-16E-10.

The Challenge. Providing students meaningful, up-to-date, and relevant practical experiences while being limited by finite resources in the provision of laboratory hardware and infrastructure.

The Solution. Equipping the laboratory with PCs, LabVIEW-based virtual instrumentation workstations, data acquisition cards, and application-specific hardware including medium power electric motor drives and electromechanical devices.

Introduction

Many academic courses have already begun incorporating computer-based educational tools for student use, either in the lectures or in the laboratory practices, or both [1]. Furthermore, information and experience sharing are becoming increasingly critical to educational institutions, as well as to practicing engineers, mainly driven by the advancements in computer technology and the Internet.

As education and technology merge, the opportunities for teaching and learning expand even more. Some forces that accelerate the changes in education include: a continual need to update and augment the content for lecture courses to keep pace with technological changes, the increasing cost of laboratory instrumentation, demands from students, and the Internet.

A number of interactive, computer-delivered simulation, control, and scientific visualization software programs are available in the market. Many application-specific tools have already been reported in the literature. These programs include Hypertext, Authorware, Director, Labtech, Visual C++, Visual Basic, Matlab/Simulink, and LabVIEW. The following criteria [1] may be used to select application software for building a virtual instrument that may be used in engineering education:

- **Modularity** — enables designers to test individual modules easily and to develop applications quickly

- **Multi-platform portability** — enables designers to work on separate parts and compile them on one platform

- **Compatibility with existing code** — allows designers to incorporate previous applications and previous versions of the software

- **Compatibility with hardware** — enables designers to gather data from different interface hardware

- **Extendable libraries** — enables designers build libraries of low-level routines to link them in higher-level systems

- **Advanced debugging features** — allows designers to optimize product design and to determine defects in the code

- **Executables** — enables designers to avoid alteration, hide the code, or create stand-alone applications

- **Add-on packages** — indicates the market acceptance of the product and speeds the development

- **Performance** — allows designers to ensure that the end product meets the required performance

- **Intuitive Graphical User Interface (GUI)** — enables users to look the screen and see what needs to be done

- **Multimedia capabilities** — available for future developments.

LabVIEW has been selected in the application described here and complies with most of the criteria described above. LabVIEW lets users create application-specific templates (sub-virtual instruments) to reduce the production time for the identical subjects. Users can incorporate with the LabVIEW programs to perform very useful tasks in a laboratory virtual instrumentation system design. Some of the functions that we used in our project are listed below:

- Adding warning sounds or messages
- Providing instructions, pre-practical tutorials, and/or interactive short-tests
- Generating a test report or data file in a common text format
- Printing a specific chart or part of the user front panel
- Linking to other currently available systems and software
- Including passwords to limit access
- Animating system or sub-system operation for easy understanding
- Providing a GUI that mimics the real instruments
- Playing a video.

The significant feature of the experimental setups used in the lab is the interfaces [2,3], which were specifically designed to operate the system safely and accurately. Furthermore, the principal objective of the laboratory development project was to provide a visual educational experience while retaining the conventional laboratory techniques. A number of educational benefits of this method include using experiment time more effectively, providing a visual educational experience, and introducing more and advanced analysis aspects.

System Architecture

The system diagram of the laboratory (Figure 5-1a) and the common features and the stages (Figure 5-1b) of the computer-assisted real-time experimental modules used in the laboratory is shown in Figure 5-1. In the laboratory, there are 10 Pentium-based PCs running Microsoft Windows NT and LabVIEW 5.1 Full Development System, 2 laser printers networked to the computers, 10 custom-built torque transducers to measure the instantaneous shaft torque in the machines, 120 custom-built current transducers (50 A and 100 A, DC to 100 kHz), 120 custom-built voltage isolation amplifiers (1000 V rms, 50 kHz), 10 benches and switchboards, 10 static starting circuits and interfaces for the slip-ring induction motors, and wiring [2,3]. Each workstation in the laboratory contains medium-power mechanically coupled rotating electrical machines: DC machine, slip-ring induction motor and synchronous machine. The static devices for the test are supplied on the bench

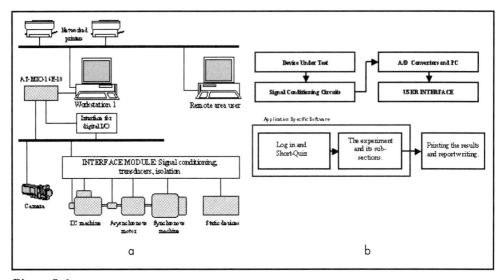

Figure 5-1
a) The system diagram of each workstation available in the laboratory. b) The hardware and the software components of each unit.

where interface terminals are available via custom-built voltage and current transducers. The accompanying hardware is highly flexible and includes suitable control signals for emergency shut down and control. We added one Logitech video camera to this system for visual communication during remote area experimenting.

Every experiment in the laboratory requires some signals to be measured on specific channels. In the system we designed, a maximum of eight parameters can be measured in real-time. This number was determined by the maximum number of parameters needed to perform the most complex experiment in the laboratory, such as three-phase voltages, three-line currents, speed and torqued. National Instruments AT-MIO-16E-10 DAQ card is used in each workstation to perform the tests, which has the following specifications: 8 differential A/D, 12-bit resolution, 100 kHz sampling frequency, 2 analog outputs (for 12-bit D/A conversion), and 8 digital I/O.

As illustrated in Figure 5–1b, after users have successfully logged in, students perform experiments in the following sequence:

- **The self-studying tutorial**—students access all required material via computer, including relevant photos.

- **The short-quiz and the experiment**—the computer asks a certain number of randomly selected "must-know" questions that cover safety issues, measurement procedures, and instrumentation details of the experiment subject. Following the successful completion of this phase, the experiment can be started. The virtual panels include the university's logo, name of the student, date and time of the experiment, and a pull-down menu indicating the subsections of the experiment.

- **Assessment and report writing**—detailed results are printed out or saved to a diskette for further analysis. The student submits a written report including the test results and the answers to the analysis questions. This report is generated at the end of the experiment by LabVIEW.

As mentioned above, each workstation accommodates a number of transducers that are used to measure high voltages and currents even at high switching frequencies (Figure 5–2). Noise immunity and personal safety are always an issue in such systems. Therefore, each workstation is designed to achieve total isolation. The voltage dividers are used to attenuate the high voltages. Using isolation amplifiers isolates the attenuated voltages. Each isolation amplifier is also powered from a separate DC/DC converter. Each group of three transducers is equipped with separate floating power supplies for additional safety. The voltage transducers are also physically guarded against potential danger that may occur due to an arc.

In order to create a buffering circuit and to take full advantage of the resolution of the A/D conversion, additional amplifiers are used to amplify the signals obtained via the Hall-effect current transducers and the isolation amplifiers. The amplified signals are transferred to BNC terminal panels via coaxial cables to eliminate unwanted signals.

Figure 5-2
Our custom-built current and voltage transducers

Sample User Interfaces

The previous system in the laboratory depended heavily on detailed written instructions and an associated tutorial session that is delivered by a lecturer. With the new method the experiments are structured in such a way (Figure 5-1a) that all the required materials are accessible via computers.

Two sample user panels are shown in Figure 5-3. The first window in the experiment, following the University's logo, asks students to enter their names. The data entered into this window are displayed at the bottom of each front panel together with the time and the date of the experiment (Figure 5-3a), and they are used as a proof of completing the test. Each experiment starts by asking a certain number of knowledge questions. A menu called Tutorial Questions links the questions. These questions are

Chapter 5 • General Test

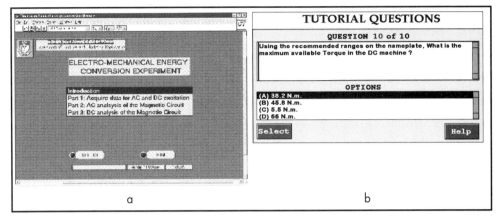

Figure 5-3
Two sample front panels (user interfaces) illustrating the opening menu of an experiment (a) and the window of the multi-choice tutorial questions (b).

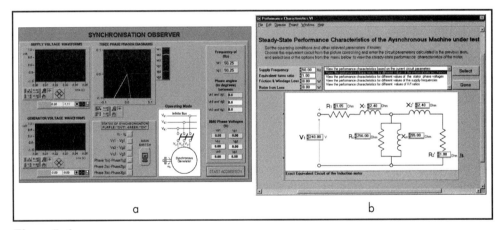

Figure 5-4
LabVIEW not only mirrors real objects or systems, but it also allows students to perform further detailed analysis based on the measured results: The front panel of the synchronization experiment (a) and the front panel of the Performance Characteristics in the asynchronous machine experiment (b).

selected randomly from a group of questions in a question bank by the personal computer allocated to each machine set.

The practical tests in the laboratory begin after successfully completing the phase illustrated in Figure 5-3. Two sample front panels of experiments are shown in Figure 5-4. The panel in Figure 5-4a is the synchronization test that requires 6 analog input channels (three-phase voltages from the synchronous machine side and three phase voltages from the supply side) to operate in realtime. Figure 5-4b shows the front panel of the performance characteristics in the asynchronous machine experiment, which accommodates various controls.

The material currently covered in our laboratory is shown in Table 5-1. The table also lists some of the experiments that are under development.

Table 5-1 Typical formal laboratory experiments and topics covered

Title	Description
Preliminary Practical Session	Discusses safety issues and measurement procedures in general
Electromagnetic Device	Demonstrates the principle operation of an electromechanical device and measurements of hysteresis loss, flux linkage and force
Characteristics and Losses of Rotating Electrical Machines	Examines the losses and measures the moment of inertia of a rotating electrical machine
Three Phase Induction Motor Tests	Determines the parameters of an exact equivalent circuit and analyses speed control principles
Synchronous Machine Tests	Determines the parameters of an exact equivalent circuit and examines speed control principles
DC-Choppers in Speed Control of DC Motors	Covers the principles of speed control using a single transistor step-down DC converter
Phase Control by Thyristor and Triac	Covers the principles of phase control at different loading conditions
Single Phase Transformer Tests	Determines the parameters of an exact equivalent circuit and examines initial switching
Real-Time Temperature Measurement and Analysis	Under development. Will analyse the impact of temperature variations on electrical circuit parameters
Synchronization	Demonstrates measurements of frequency, rms, and phase sequence on three-phase power lines
Principles of Phasors and Rotating Field in AC Machines	Demonstrates the likeness between the three-phase real-time signals and phasors at different loading conditions and the concept of rotating fields

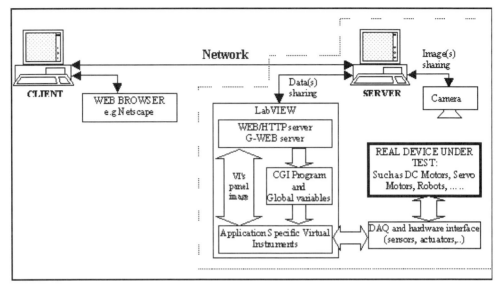

Figure 5-5
The components of the remote experiment system in the LabVIEW implementation

Remote Area Experiments over the Internet

Remote area experimenting and instrument control over the Internet is becoming commonplace in engineering education. The common feature of this practice is to increase the utilization of unique resources by enhancing access. While implementing such systems, it should be noted that the plants used in remote experimentation should have a full local control but limited remote control. Furthermore, due to the existing speed limitations of the Web access, a virtual representation of the moving part in the experiments should be used instead of the real image.

To publish virtual instruments on the Web and remotely control them from any Web browser, you must have the LabVIEW Internet Developers Toolkit and know about URL, G Web Server, HTML, and Common Gateway Interface (CGI). The block diagram given in Figure 5-5 shows the basic components of the remote experiment system in the LabVIEW implementation, which is also used in our pilot system.

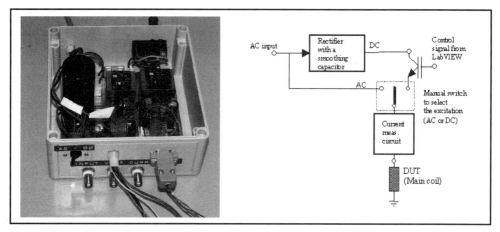

Figure 5-6
The custom-built signal conditioning circuit and the block diagram for the remote area experiment

A pilot experiment on electromechanical devices was selected to try out remote area experimenting in the laboratory. After successfully testing the method, we have now decided to develop a robust hardware for this experiment. The photo of the custom-built signal conditioning device and its block diagram is shown in Figure 5-6. This setup can provide complete isolation and can also allow the user to perform AC or DC tests on the device under test.

Results

The roles of teachers and students are changing, and there are undoubtedly ways of learning not yet discovered. However, the computer and software technology may play a significant role in identifying problems, presenting solutions, and encourage lifelong learning. Furthermore, it is obvious that many things that we are doing can be done better with the help of technology. However, the selection criteria for software are still major issues.

The Internet has the potential to provide a highly supportive learning environment and may attract new students and add value to education.

As reported in the paper, an interactive, computer-controlled, highly flexible test system also suitable for remote area experiments has been designed and implemented successfully. The key benefits of the system are:

- Elimination of prepractical tutorials
- Less time in each session (means more sessions to handle the large number of students and/or advance study)
- Saved resources (student time and demonstrator time)
- Accurate measurement for detailed and dynamic analysis
- Remote access via the Internet

The system described here has been developed and implemented successfully. The laboratory has been functional since the first semester of 1999 and, in total, will serve approximately 400 students each semester. Our plan is to increase its capabilities of remote area experimentation and increase the usage of visual and audio communication through the Internet for distance experimenting.

References

[1] N. Ertugrul, "Towards Virtual Laboratories: A Survey of LabVIEW-Based Teaching/Learning Tools and Future Trends," *International Journal of Engineering Education*, Special Issue on LabVIEW Applications in Engineering Education, expected to be published in early 2000.

[2] N. Ertugrul, A. P. Parker, and M. J. Gibbard, "Interactive Computer-Based Electrical Machines and Drives Tests in the Undergraduate Laboratory at The University of Adelaide," EPE'97, 7th European Conference on Power Electronics and Application, Trondheim, Norway, 8-10 September 1997.

[3] N. Ertugrul, "New Era in Engineering Experiments: An Integrated Interactive Teaching/Learning Approach and Real Time Visualisations," *International Journal of Engineering Education*, Vol.14, No.5, pp. 344-355, 1998.

■ Contact Information

Nesimi Ertugrul
        Department of Electrical and Electronic Engineering
        University of Adelaide
        Adelaide, 5005, Australia
        Tel: 61-8-8303 5465
        Fax: 61-8-8303 4360
        E-mail: nesimi@eleceng.adelaide.edu.au

PC-Based Data Acquisition System for Measurement and Control of an Isotopic Exchange Installation

Retevoi Carmen Maria
Engineer
National Research Institute for Cryogenic and Isotope Separation
Romania

Stefan Ovidiu Liviu
Engineer
National Research Institute for Cryogenic and Isotope Separation
Romania

Cristescu Ion
Manager of the Cryogenic Pilot Plant
National Research Institute for Cryogenic and Isotope Separation
Romania

Products Used. Interface AT-MIO-16-XE-10 data acquisition,, SCXI-1100 signal conditioning module, LabVIEW.

The Challenge. Developing a data acquisition system for a catalyst isotopic exchange module from a pilot plant for tritium and deuterium separation.

The Solution. Designing a virtual instrument to acquire the signals from temperature sensors of catalyst isotopic exchange module and to analyze and command the flow rate and power supply for electrical heaters. We performed the multi-channel acquisition with the SCXI 1100 and AT-MIO-16-XE-10 from National Instruments. We chose signal conditioning because of the following advantages: electrical isolation, transducers interfacing, signal amplification, filtering and high-speed channel multiplexing.

Introduction

CANDU is one of the most performed technologies for nuclear power plants. CANDU-type reactors use heavy water as a moderator. After a period of functioning, the deuterium concentration decreases and tritium concentration increases. The increase of tritium concentrations in heavy water determines some problems in the operation of a nuclear reactor and also for the environment. Therefore, it is very important to decrease the tritium level in heavy water. Many owners of CANDU reactors are conducting research and developing technologies for heavy water detritiation.

In Romania, we have a nuclear power plant with a CANDU-type reactor. After several years of operation, the radioactive level in the moderator attained such a value that heavy water detritiation began to occur. Therefore, the Institute of Cryogenic and Isotope Separation is developing a research program for tritium separation from heavy water. For this purpose, we established an experimental facility called the Pilot Plant for Tritium and Deuterium Separation. One of the plant's components is the catalyst isotopic exchange module, where tritium is transferred from a liquid phase to a gaseous phase.

The functioning principle of the module is catalytic isotopic exchange between heavy water and hydrogen gas into an exchange column. The heavy water is heated in an electrical heat exchanger for raising temperature at 60-90 °C, and then it is introduced in the exchange column. Hydrogen gas is also warmed in an electrical heat exchanger at 70-90 °C and then is introduced in the exchange column. In this way the exchange column is controlled at a fixed temperature. The temperatures for hydrogen gas, heavy water and the exchange column are measured with thermocouples and platinum PT100, and the measuring signals are introduced into the monitoring system. To maintain a constant column temperature, the analysis and monitoring system commands a power supply for the electrical heat exchangers. Other parameters measured in the flow circuit are the heavy water level in the exchange column and storage vessels. Signals for these parameters are also monitored by an analysis and monitoring system.

We used a personal computer, LabVIEW software, a data acquisition board, and SCXI-1100 signal conditioning to acquire, control, and display test data for isotopic exchange installation from the Cryogenic Pilot Plant (Figure 5-7).

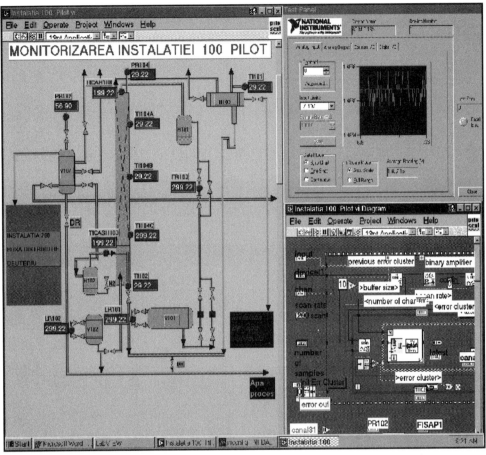

Figure 5-7
Monitoring the catalyst isotopic exchange installation from the Cryogenic Pilot Plant

Product Development Performances

Because of the number of parameters (temperatures, pressures, and levels) in the installation, we chose the SCXI-1100 with 32 differential channels. The selectable gain and bandwidth settings are ideal for configuring the module to condition a variety of millivolt and volt signals. The SCXI module multiplexes the 32 channels into a single channel of the AT-MIO-16XE-10.

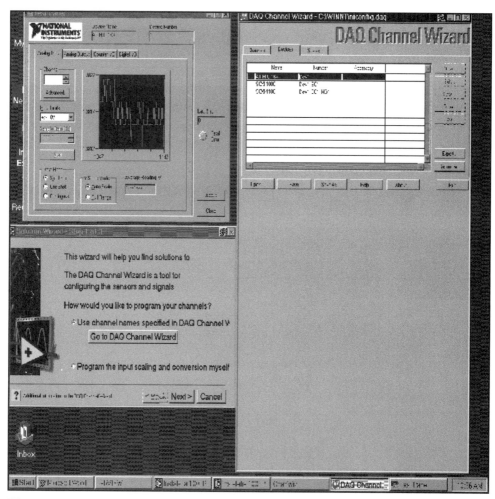

Figure 5-8
DAQ Solution Wizard for configuring the sensors and signals connected to the AT-MIO-16XE-10 DAQ board and the SCXI-1100

We used SCXI to electrically isolate the sensors and the monitored system from the data acquisition system and the host computer. We leave the jumpers W5, W2, and W9 in their factory-default position. All of the sources are floating, so we configured W1 to connect a 100 kΩ resistor to the negative input of the amplifier to prevent saturation of the amplifier inputs.

The DAQ Solution Wizard guides us through naming and configuring the sensors and signals connected to the data acquisition board (Figure 5-8).

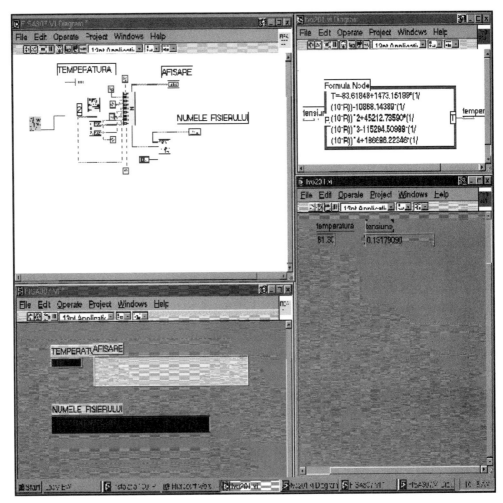

Figure 5-9
Data saved in LabVIEW

The NI-DAQ product line addressed by a LabVIEW source code allows us to develop an easy-to-use, low cost system.

Results

This solution performs a large number of acquisitions and achieves many advantages from SCXI.

The gain is applied to the low-level signals from the SCXI module located close to the transducers, sending only high-level signals to the PC and minimizing the effects of noise on the readings.

All data acquired are saved in specified files and are used to determine the thermodynamic parameters associated with each measurement point (entropy, enthalpy, specific heat, and liquid fraction), as shown in Figure 5-9.

With these values, we analyze the performances for the compounding elements (heat exchangers, turboexpanders, and throttle valves) of the cyogenic cycle using the temperature-entropy diagram.

We used LabVIEW because it is very simple to chart, analyze, and save data. LabVIEW has given thousands of successful users a faster way to program instrumentation and data acquisition systems. By using LabVIEW to prototype, design, test and implement, we reduced system development time and increased productivity.

■ Contact Information

Retevoi Carmen Maria
    Engineer
    National Research Institute for Cryogenic and Isotope Separation
    CP10, Rm. Valcea
    Romania 1000
    Tel: 0040-050-732744
    Fax: 0040-050-732746
    E-mail: retevoi@usa.net

LabVIEW Tests M1A1 Ammunition

Lance Butler
Senior Systems Integrator
B & B Technologies, Inc.

Products Used.
National Instruments Hardware. PXI-1000 8-slot PXI chassis,
PXI 8150 233 MHz PXI controller 128Mb RAM with GPIB,
PXI-8210 Fast Ethernet and Wide Ultra SCSI Interface,
2 NI-5102 PXI digital oscilloscope, PXI 1408 IMAQ card,
5 FP-AI-110 FieldPoint 8 channel 16 bit analog input module,
FP-1000 RS-232/RS-485 FieldPoint network module, and
4 FP-1001 RS-485 FieldPoint Network Module.

National Instruments Software. LabVIEW 5.1, NI-DAQ, FieldPoint Server, SQL Toolkit, NI-IMAQ™, IMAQ Vision, and NI-GPIB.

Third-Party Hardware. Black Box Industrial RF modems 5 ea. and Weibel Radar .

The Challenge. Developing an automated testing system capable of remote measurement of various aspects of the Tank Ammunition Testing process, including wind speed and direction, target impact location, projectile velocity, and muzzle pressure.

The Solution. Using LabVIEW and PXI were used to create a measurement system capable of measuring all aspects of the firing of the rounds and reporting the data to an Microsoft Access database.

Introduction

Alliant Techsystems, Inc. is the leading manufacturer of M1A1 tank ammunition for the U.S. Army, including a variety of 120 mm rounds (Figure 5-10). Various acceptance tests must be done for each lot of ammunition parts before the government will accept them. As part of this testing procedure, they fire approximately 40 rounds of ammunition per testing day to prove that the rounds will perform when it counts. When it came time to upgrade the system, LabVIEW and PXI were the obvious choice for an efficient,

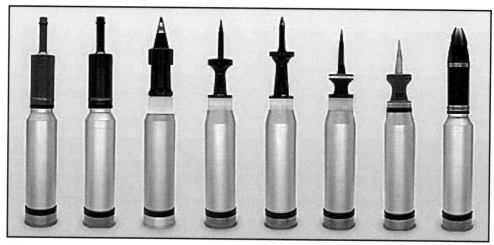

Figure 5-10
Alliant produces an entire family of tactical and training 120 mm rounds that are interoperable and interchangeable among NATO 120 mm smoothbore tank forces

upgradeable system. Alliant Techsystems called B & B Technologies, Inc. to design and integrate the new system.

Testing Procedure

The system consists of three tank cannons that can be used for each test. Two of the cannons are used for short range (200-300 meters) shots where pressures in the barrel and muzzle exit times are the key data. The third gun is used for long range firing (3 kilometers) where accuracy of impact is also recorded.

Each test day begins with a firing order generated in Microsoft Access. LabVIEW monitors this database through Structured Query Language (SQL) and reads the firing order into the graphical user interface (GUI) once it is created. Each successive shot is then armed from LabVIEW and fired by the gunner.

The rounds are fired through a current applied to the primer. Two pressure transducers in the barrel are fed into one of the oscilloscope cards to record the pressure curves (around 150,000 PSI) that propel the bullet. A second oscilloscope card is used to record the pulse that fires the round, as well as, the pulse generated by a flash detector that triggers when the muzzle exit

Chapter 5 • General Test

Figure 5-11
M1A1 cannon firing test round

flash (a 10-meter diameter fireball) appears outside the barrel. A Weibel radar unit is controlled through GPIB to determine the velocity of the round when it exits the barrel. Downrange cameras transmit video data through a microwave link back to the image acquisition (IMAQ) card in the bunker. Five seconds of flight video data are recorded for later playback. FieldPoint modules read voltages off of wind speed and direction transducers. The data is then shipped back to the FP-1000 in the bunker through wireless modems. One FieldPoint module at the bunker is used to record ambient temperature and pressure.

After the shot, the operator verifies that the oscilloscope data does not indicate any problems with the transducers (150,000 PSI is hard on cables and transducers). Next, the video data is used to pinpoint the impact point of the round in the target. Finally, the operator tells the LabVIEW program to export the data to Microsoft Access. Various calculations are made on the raw data and the results are passed through Microsoft Access. The data is all zipped at the end of the day and transmitted through email from the test facility in New Mexico to the engineers in Minnesota for review.

Figure 5-12
This in-flight photo shows the M829, an armor-piercing, fin-stabilized, discarding sabot tracer round with a depleted uranium penetrator

■ Contact Information

Lance Butler
Senior Systems Integrator
B & B Technologies, Inc.
6610 Gulton Court NE
Albuquerque, NM 87109
Tel: (505) 345-9499
Fax: (505) 345-9699
E-mail: lbutler@bbtechno.com

Industrial X-Ray Digital Image Flaw Detection System

XiaoMing Kou
Manager, Vision Department
Shaanxi Hitech Electronic Company, Ltd.

EnQuan Guo
Shaanxi Hitech Electronic Company, Ltd.

Xingfen Zhao
Shaanxi Hitech Electronic Company, Ltd.

Products Used. LabVIEW 5.0, IMAQ Vision, SQL Toolkit for G2.0, IMAQ PCI-1408.

The Challenge. Updating an existing X-ray industrial television and conventional X-ray film photographic inspection system to a digital image flaw detection system, which leverages off the methods of digital image processing and a database management system so that advanced performance can be obtained.

The Solution. Developing the Industrial X-ray Digital Image Flaw Detection system using IMAQ Vision, a PCI-1408 image acquisition board, and LabVIEW to create a system which costs less and does more in less time than competing solutions.

Introduction

X-ray industrial television is a nondestructive test method, which uses a television, cathode ray tube (CRT), and X-ray film to inspect flaws in products. There are several disadvantages of using this method:

- Taking X-rays requires a sequence of exposure, developing, and fixation. This process is expensive, takes a long time, and cannot be graded real-time.

- It is difficult to measure the flaw on a CRT, so the user cannot quantitatively analyze the flaw.
- Developers cannot take advantage of the database management system to manage the flawed data and pictures.

Companies recognize that quality is important. Industrial vision plays a key role in meeting the quality challenge. With the development of the PC, a new term has become popular in the instrumentation community, virtual instruments (VIs) — this term describes the combination of programmable instruments with general-purpose PCs.

In order to use the technique of image acquisition and processing, as well as the database management system, we developed a new kind of nondestructive test instrument to meet these needs.

System Hardware

The host computer for this system is a Pentium II 400 MHz with 128 MB RAM, a 6.5 G Ultra ATA 7200 RPM hard drive, and a HP CD-Writer Plus 8100I running Windows 98. We chose a 17-inch CRT with a 1:1 aspect ratio so the function panels do not cover the X-ray image windows and we can still measure the flaws easily.

The X-ray source is stretched into a steel pipe and aimed at welded joints. A filter and image enhancer accepts the X-ray in the corresponding position outside the pipe. The X-ray penetrating through the welded joints is transformed into visible optics and passed to a charge-coupled device sensor. This composite video signal is acquired by the PCI-1408 and transformed into a digital image that can be viewed on a PC.

System Software

We chose LabVIEW as our development environment because with it we can rapidly generate a graphical user interface for display, analysis, and control. For image processing and analysis we used IMAQ Vision, which is a LabVIEW toolkit with many kinds of image processing tools and functions.

We used Microsoft Visual FoxPro 6.0 to store information on the flaws that the system found. Each record in this database has sixteen fields, including record number, picture name, flaw classification, flaw parameter, flaw location, product class, and weld grading information.

In order to access the FoxPro database in LabVIEW, we used the SQL Toolkit for G2.0. It has the open database connectivity interface needed to connect LabVIEW and the database. For printing reports, we used Borland C++. This program is designed as an executable program with LabVIEW. We used BDE Administrator 5.0 to do this.

System Function and Performance

This system has the following functions:

- Replaces conventional X-ray film photographic inspection. Conventional X-ray inspection costs more, takes more time, and cannot be graded real-time.

- Includes digital image acquisition and processing functions. Because we used the PCI-1408 to acquire X-ray television signals and convert them into digital images, we can use digital image processing functions. The X-ray image takes only part of the field of visibility. We use a lookup table to set the contrast and the brightness of the image. The flaw is gray and neither black or white. In order to increase the contrast, we designed an S arithmetic operator which enlarges the gray grade around the flaw and reduces the gray grade near the black and white. We use the spatial filter and edge enhancement to make the inspection effective. We also use morphology and an analysis function to analyze the flaws in the image.

- Allows the operator to discriminate between flaws and classify them as inadequate penetration, incomplete fusion, internal concavity, burn-through, slag inclusion, porosity, etc. with image processing.

- Allows the operator to measure flaw parameters in the image window, append flaw information to the database, save the flaw picture, and record the information.

- Includes capability for the database to be browsed and edited.

- Supports online help.

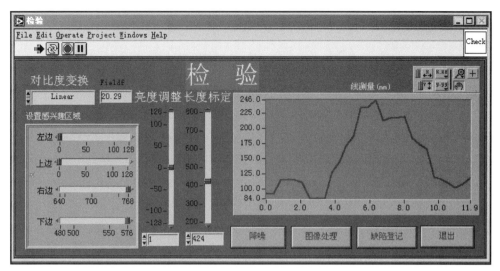

Figure 5–13
Inspection panel of the Industrial X-ray Digital Image Flaw Detection system

The main menu of the Industrial X-ray Digital Image Flaw Detection system includes login, check, browse, print, system maintenance, and online help. Figure 5–13 shows the inspection panel.

In the inspection panel, the user can set the region of interest and contrast information for the image acquisition board in order to save time. When the operator finds a flaw point, a filter function is initiated. This results in a static image. The operator uses a line profile to measure the flaw in the line measure window. If the flaw is difficult to see, the operator can use image processing to see the flaw clearly. Once the flaw is logged in, a record for the flaw is added to the flaw database. Then the program goes into dynamic acquisition automatically if the quit button is not pushed.

The performance of this nondestructive test system can be summarized as the following:

- In the Phase Alternate Line (PAL) format, image acquisition can get up to 768 x 576 pixels and display in 25 frames/second

- Static sensitivity is higher than 2 percent

- Dynamic sensitivity is higher than 3 percent

- Streaming image-to-disk can be made up to 12.5 frames/second.

Results

Based on LabVIEW, an image acquisition board, and IMAQ Vision, we developed the Industrial X-ray Digital Image Flaw Detection system rapidly and with more functions and higher performance than the conventional X-ray film photographic inspection system. This nondestructive test system can be widely used to inspect the internal quality of metal, and weld seams of pipe.

■ Contact Information

XiaoMing Kou

Manager, Vision Department
Shaanxi Hitech Electronic Company, Ltd.
No. 18, Gaoxin Road
Xi'an, China
Tel: 8629-8311231
Fax: 8629-8237106
E-mail: Hitbb@public.xa.sn.cn

Large Area Conditioning Systems for the National Ignition Facility

Richard Jennings
Technician
Sandia National Laboratories

Products Used. LabVIEW 5.0, IMAQ Vision 4.1.1, IMAQ PCI-1424, IMAQ PCI-1408, PCI-GPIB, and PCI-MIO-16XE-50.

The Challenge. Integrating multiple pieces of R&D laboratory equipment and computer systems into one easy-to-use tool.

The Solution. Using LabVIEW 5.0 and the multithreading capability to perform simultaneous image acquisition, data acquisition, and motion control using one off-the-shelf PC with dual 400 MHz Pentium II processors.

Introduction

The National Ignition Facility (NIF) is a 192-beam laser system that will require the delivery of over 1,200 high damage threshold, meter-class, coated optical components before completion. The NIF optics group at Lawerence Livermore National Laboratory (LLNL) was able to increase the damage threshold of coatings placed on mirrors and polarizers by developing a process to condition each optic after it comes out of the coating chamber. The optic is mounted in a custom translation stage and the coated surface is scanned in a raster pattern through the waist of a high-power Nd:YAG laser running at 30 Hz. Continuous monitoring of coating quality and the process laser beam need to be performed in parallel with motion control. To meet these requirements, LLNL built the Large Area Conditioning system (LAC) for continuous operation at the coating vendors' facilities (see Figure 5-16).

The technical requirements for the LAC were well defined because the process had already been proven on a prototype system. Our goal was to integrate multiple pieces of equipment and the test procedure into one easy-to-use system capable of unattended operation. Management's vision was of a fully automated test system with one or two buttons, capable of being run by a moderately skilled technician.

Development

Our development machine was a Compaq AP400 with dual 400 MHz PII processors and 128 MB RAM running Microsoft Windows NT 4.0. LabVIEW 5.0 was recently released with support for multithreading, a feature which allows concurrent execution of processes on computers with more than one central processing unit (CPU). A dual-processor PC running Microsoft Windows NT 4 provided us with an opportunity to try out LabVIEW in a multithreading, multitasking, parallel processing environment.

Our first task was to break down the largely manual prototype LAC into functional components and decide how each component could be optimized and how it would interact with the others. We decided to break the system down into three different components: the beam profiler, test code, and the user interface. Development of the beam profiler and the test code followed similar paths. We programmed the hardware to perform buffered acquisition directly to host RAM each time an external trigger occurs, processing the data as it becomes available. The user interface reads data from functional globals and displays it at 2 Hz. Each component was developed and optimized independently before the system was integrated.

We used Microsoft Windows NT Task Manager (Figure 5–14) to give us an idea of how efficiently our code was using the processors and memory available to us. The LabVIEW internal profiler was critical for seeing how much time LabVIEW spent executing a virtual instruments (VI) code, but more importantly the graphical display of the task manager allowed us to visualize how changes in program structure affected the load on the computer. We wanted to make sure that no task would be starved for CPU cycles in the final product, so we set a development goal that at no time during execution would any single component use more than 50 percent of the processors. In the final product, our total processor usage peaked at ninety-five percent.

We were able to cut down on the time Microsoft Windows NT spent swapping data in and out of virtual memory by preallocating arrays and image buffers. Setting up all of our data acquisition and image acquisition to use circular buffers allowed us to perform other tasks while monitoring the buffers for new data. If the code processing the data in the circular buffer is delayed while other tasks are executing, it can catch up later as long as the

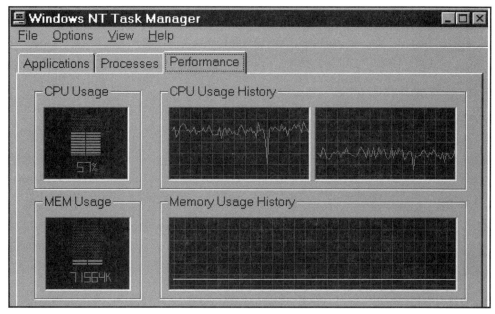

Figure 5-14
Microsoft Windows NT Task Manager, which monitors CPU and memory usage

data has not been overwritten. This feature turned out to be a key to success later on when the different components were integrated.

Beam Profiler

Precise measurement and control of the energy in the process laser is critical to laser conditioning of the coating—too much energy and the coating could be destroyed, too little energy and the coating wouldn't survive on the NIF laser. We were able to deliver a self-calibrating instrument that ran faster than its predecessor by building our own integrated beam profiling system as a virtual instrument within LabVIEW using National Instruments digital framegrabber, the IMAQ PCI-1424, and the Pulnix TM 9701 digital camera.

A partially reflective optic was placed into the beam path of the conditioning laser beam and a small portion of the beam was redirected to our diagnostics and analyzed for energy and intensity. The Pulnix digital camera and the PCI-1424 were configured to capture an image with each laser trigger. The laser initially ran at 30.0 Hz but the camera had problems synchronizing to that frequency. Turning the laser's repetition rate down to 29.97 Hz allowed us to trigger the camera in a progressive scan mode at video rate and capture a full 768 V x 484 H frame of every shot.

NI-IMAQ, National Instruments device driver for the IMAQ PCI framegrabber, has several different acquisition modes. We chose to set up a circular buffer 10 frames deep in the host RAM and have the driver place an image of each laser shot sequentially into the buffer. When the end of the buffer is reached, the driver overwrites older images at the beginning of the buffer with newer images. Acquisition continues in this circular manner until the program is stopped.

Image acquisition occurs at the driver level while the higher-level analysis routine only needs to monitor the image buffers for new data without being directly involved in the acquisition details. If the analysis routine is delayed from execution by some other task, the last 10 images are always in memory so it can easily catch up as long as the delay has not been longer than 333 milliseconds (10 images). The commercial beam profiler in our prototype system would capture an average of one shot per second; in the final product our beam profiler could go over one million shots without missing a shot.

Analog Data Acquisition

Coating quality is monitored by illuminating the surface of the coated optic with a laser diode and measuring the amount of back-scattered light with a photodiode mounted on a customized microscope system. Defects in the coating cause more light to be scattered back to the photodiode, which generates a DC voltage proportional to the amount of scattered light. This voltage is measured before and after each shot of the 30 Hz Nd:YAG laser with a National Instruments PCI-MIO-16XE-50 DAQ card and analyzed in LabVIEW.

NI-DAQ, National Instruments device driver for their data acquisition cards, allowed us to use a pre-trigger synchronized to the process Nd:YAG laser as an external scan clock for our data acquisition. The channel order and delay between samples was configured so that each time the DAQ card received a trigger it would measure the photodiode approximately 500 microseconds before and after the laser shot. If the difference in scattered light from before the shot and after the shot is above a predetermined threshold, the process is stopped before the next laser shot can cause further damage. Because data acquisition occurs at the driver level just like image acquisition, the higher-level analysis routine is free to operate without being directly involved in the acquisition details. DAQ occurrences were used within LabVIEW to notify the analysis routine that data was available.

One of the requirements of the LAC was that the entire coated surface of the optic be scanned through the process laser beam at a constant velocity without multiple shots from the process laser occurring on top of each other. The motion controller was configured to scan past the test area at the beginning and ending of each scan line. This over scan region allowed the motion controller to accelerate and decelerate the optic while keeping a constant scan velocity during the test. The motion controller set a digital line high only when the coating was in the correct position. This digital line was used to turn on and off a shutter for the laser, as well as gate the data acquisition. When the entire surface of the coating had been conditioned, areas that showed a significant change during the test are photographed with a CCD on the microscope and National Instruments IMAQ PCI-1408 analog framegrabber.

User Interface and Execution Systems

Since the beam profiler and the data acquisition were both hardware timed to an external event, we needed to implement our user interface without interfering with these hardware timed processes. We decided to implement our program as three separate parallel loops on the main diagram with communication between the loops using global circular buffers. The user interface ran at 2 Hz reading data from the buffers for display. Once the main program was started and all the systems initialized, execution of the three separate loops could take place in parallel without interfering with each other.

Earlier versions of LabVIEW used a timeslicing technique to give the appearance of parallel execution, but it was not until LabVIEW 5.0 introduced multithreading that we were truly able to have VIs running in parallel on multiple processors. LabVIEW 5.0 took care of all the details of splitting our code up into threads for parallel execution on both processors. As developers, we had to determine whether to assign execution systems and priorities to each section of code or to let LabVIEW use its default settings. We chose to leave the priorities at the default setting but assign separate execution systems to the top-level VI of the beam profiler and the data

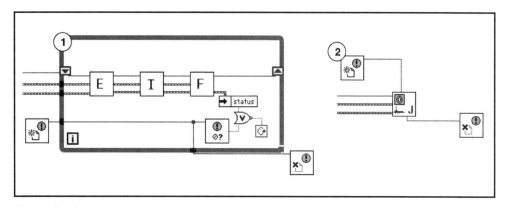

Figure 5-15
The while loop (1) runs at 30 Hz. Sub-VIs E, I, and F are in a different execution system than the calling VI, resulting in 180 context switches/second. The solution (2) was to make a sub-VI, that ran continuously and never returned to the caller. Note the use of notifiers to terminate execution.

acquisition. All sub-VIs were set to run in the same execution system as the caller.

Context switching occurs when the operating system has to switch between execution systems. There is some overhead involved when context switching occurs, and this was noticeable during the development phase of the beam profiler. The initial beam profiler had three main VIs in a while loop located on the main VI diagram. The beam profiler ran in a separate execution system from the main VI, so every time this loop executed the operating system (OS) would perform six context switches, one as each VI in the loop was called and again when each VI returned. Since the beam profiler ran at 30 Hz, this meant we were forcing the OS to perform 180 context switches a second. The solution was to turn the while loop with the beam profiler VIs into a sub-VI that was called once and never returned to the main VI (Figure 5-15). This strategy allowed our beam profiler and the rest of the system to run in parallel without interruptions.

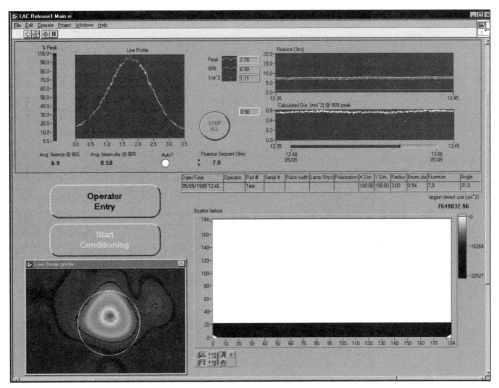

Figure 5-16
Large Area Conditioning system (LAC) front panel

Results

Careful construction of our application allowed us to maximize the ability of LabVIEW 5.0 to run code in parallel on a dual-processor PC. Acquiring our data into circular buffers each time an external trigger occurred gave us a system that was flexible enough to multitask without missing critical data. Rigorous testing of our code to ensure that no single task dominated the host PC during development gave us a final product that was responsive and tolerant of demands put on it by the user interface.

■ Contact Information

Richard Jennings
Technician
Sandia National Laboratories
7011 East Ave. MS 9055
Livermore, CA 94550
Tel: (925) 294-2696
Fax: (925) 294-2550
E-mail: rtjenni@sandia.gov

Controlling Aeronautical Hydraulic Actuator Testing with LabVIEW

R. Kyle Schmidt
Control Systems Specialist
Messier-Dowty Inc.

Products Used. PCI-MIO16XE-10, SCXI-1000, SCXI-1121, SCXI-1321, SCXI-1161, and LabVIEW 5.1 Professional Development System (PDS).

The Challenge. Controlling the endurance testing of hydraulic linear actuators for aircraft.

The Solution. Using LabVIEW, data acquisition (DAQ), signal conditioning (SCXI), and a custom SCXI module to enhance the accuracy of the testing and increase the test speed.

Introduction

The endurance testing of linear hydraulic actuators involves sequencing the actuator from extension to retraction and back while applying a force along the actuator's axis. The forces are usually specified as a function of actuator position. While many commercial systems are available to perform force control for materials testing, very few commercial products offer the capability of specifying forces as a function of test article position. The few that offer the capability are exceptionally expensive for single actuator testing. The majority of commodity off-the-shelf (COTS) force controllers only provide force definition as a function of time. Attempts to use COTS controllers with force as a function of time control to test actuators result in large amounts of time spent tuning the system for each minor change in test setup.

The solution was to integrate off-the-shelf National Instruments hardware with a custom-designed SCXI module. SCXI forms the backbone of the signal conditioning used in our lab. LabVIEW-based software allows a simple interface to the custom module (Figure 5–17).

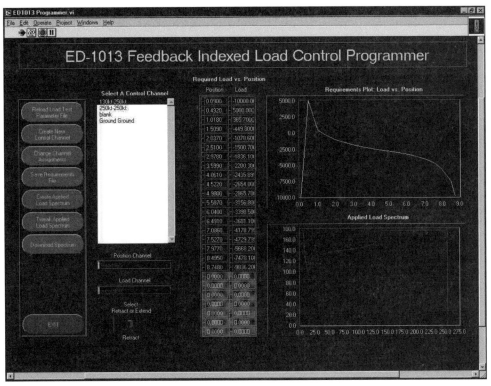

Figure 5-17
LabVIEW interface to custom SCXI module

Custom SCXI Module

The custom SCXI module is a microcontroller-based hybrid analog/digital controller. The ED-1013 module (Figure 5–18) uses the signal conditioning of a neighboring SCXI-1121 module. The conditioned analog signals are routed between the modules by using SCXI-1351 one slot cable extenders connecting the 50 pin MIO connectors. Two analog signals are used: a position input and a force feedback input. The ED-1013 uses an analog-to-digital converter to read the position input. The onboard microcontroller compares that value to an internal look-up table. The value in the look-up table is output to a digital-to-analog converter. This command value then enters the analog proportional integral derivative (PID) control loop. The PID terms are controlled by

Figure 5-18
ED-1013 custom SCXI module

other voltages output by the digital-to-analog converter. The resulting drive signal is sent to a current buffer that drives an electrohydraulic servo valve. This valve controls the flow of hydraulic fluid to the loading actuator. Other features are included in the module to allow external command generation, multiple look-up tables, and error checking.

The module is based on two microcontrollers — one to handle SCXI bus communication, the other to manage the actual control functions. SCXI bus digital communication uses the SPI protocol, and many digital logic chips and microcontrollers are available to support it. The communications microcontroller stores, then programs the test controller. All force levels and PID tuning parameters are stored in non-volatile memory (E^2PROM).

The interface to the custom module is through two high-level LabVIEW virtual instruments (VI). The VIs communicate with the module using calls to the NIDAQ32.dll shared library. Bidirectional digital communication is supported and used for the checking of PID loop gains and stored force setpoint conditions. The VI integrates with Messier-Dowty's data acquisition system, MDAQ-T. With the MDAQ-T system, a user defines a test with a test parameter file, which stores all the pertinent paths and parameters for any given test. In addition, it provides a pointer to the configuration file which stores all the calibration data for the instruments in the test. The programmer VI loads the test parameter file and presents to the user a list of virtual control channels which contain different force profiles and different configurations. Multiple control channels may exist for a single controller, but only one may be stored on the ED-1013 at any given time. Once the user has entered the programmer VI, defining a control channel is as simple as indicating which channels (from the MDAQ-T configuration) will be used for force and position feedback. The user enters the required force conditions (either manually, or loaded from an ASCII file). The programmer VI then calculates the appropriate binary values to be downloaded to the ED-1013 and stores these with the virtual channel information. The user has the option to download the control channel information to the ED-1013. The programmer, by way of the virtual control channels, allows for nearly unlimited numbers of controllers and channels.

Once a control channel is downloaded, the user may tune (at any time) a controller by using the tuning parameters VI (Figure 5–19). The tuning parameters VI interrogates the module at VI startup and sets its front panel controls to represent the current state of the ED-1013. When the update control is pressed, the VI only sets those values which have changed (to save on the number of writes to the controller E^2PROM).

System Layout

The actuator under test (AUT) is mounted in a test frame (Figure 5–20) such that its rod is joined to the rod of the loading actuator by means of a load cell. Separate hydraulic systems power the actuators. A directional control valve sequences the AUT. This valve is selected by using an SCXI-1161 power relay module. The loading actuator is controlled by means of an electrohydraulic servo valve. This is controlled by the ED-1013. A LabVIEW-based program sequences and controls the testing (Figure 5–21). Once the user has programmed the ED-1013, the module runs stand-alone — it should only be accessed to update the look-up tables or to change the loop tuning parameters.

Figure 5-19
Tuning parameters VI

The LabVIEW VI that sequences the test uses only digital input and output (IO)—discrete outputs to control the solenoid valves, and discrete inputs to sense endpoints. As the controller is running only digital IO, analog input operations are run concurrently, allowing continual viewing of data and periodic storage of data to disk. The controller runs as a plug-in module to MDAQ-T (Figure 5-22). MDAQ-T allows multiple users to run multiple tests using a single test station and appropriate signal conditioning. Since the controller runs as a plug-in, it acquires and saves data in an MDAQ-T compatible (datalog) format. MDAQ-T data files are automatically stored in time and date coded directories and files, and are compatible from one station to the next. This facilitates retrieval of archived information and the production of reports.

Chapter 5 • General Test

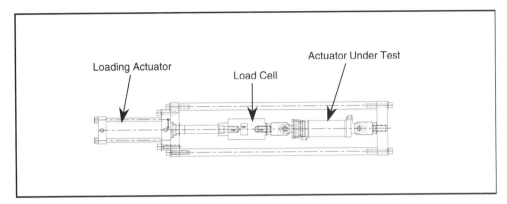

Figure 5-20
Test rig layout

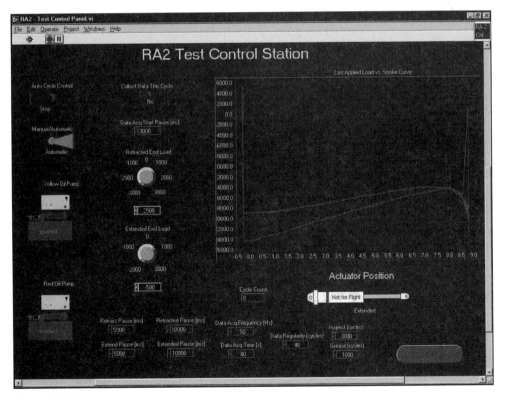

Figure 5-21
LabVIEW actuator test controller

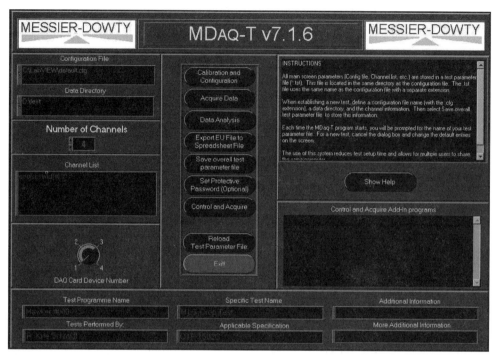

Figure 5-22
MDAQ-T main panel

Benefits

The use of this system has numerous benefits. National Instruments DAQ and SCXI products are the mainstay of our Engineering Test facility. Their modularity provides a cost-effective method to deal with a rapidly changing test environment. Individual SCXI modules allow systems to be exceptionally scalable from a four-channel test to a forty-channel test very rapidly. Traditional methods of performing actuator endurance tests have involved the use of materials testing systems to control the applied force. This results in poor performance (as the controller cannot adapt to changes in the rate of actuator motion) and high cost (single channel force controllers with programmable spectra cost in the neighborhood of $20,000). Added to this was the requirement for data acquisition—for which our lab was already using SCXI and E series DAQ boards. The ED-1013 reduces test complexity by removing the expensive external controllers, by reducing the need for redundant instrumentation (some for data acquisition and some for the controller) and by being designed to perform any force versus position control.

In addition to the reduction in complexity comes an increase in the accuracy of applied force. Materials testing systems are generally programmed with force as a function of time—should the actuator speed change from the baseline, the applied force goes out of sync with the actuator's motion—resulting in large errors. The ED-1013 eliminates this by applying the appropriate force at the appropriate position. This is particularly significant when performing temperature varying endurance tests—the AUT moves quickly at high temperatures and slowly at low temperatures. With legacy systems, considerable downtime was experienced while the legacy system was retuned to compensate for the change of AUT speed. The ED-1013/SCXI solution transitions seamlessly between low and high temperature testing.

The ED-1013/SCXI system is readily scalable—the next application in our lab will involve a single test using ten ED-1013s to control simulated airload on aircraft landing gear. Approximately forty analog data channels and thirty-two digital data channels will be simultaneously acquired.

ED-1013 modules cost less than $3000, a bargain compared to material testing controllers. The combination of National Instruments DAQ and SCXI allows a complete test control and data acquisition package to be priced less than half the cost of the off-the-shelf controllers for a single channel.

Results

The ED-1013/SCXI combination provides an inexpensive way to perform actuator endurance testing. The combination is also scalable to multiple channel endurance tests. The use of LabVIEW-based soft controllers and SCXI digital input/relay output modules allows a test stand to be rapidly setup and configured. Combining control and data acquisition in one package increases reliability and eliminates redundant instrumentation. The overall combination can reduce cost by one-half and test time by one-third.

■ Contact Information

R. Kyle Schmidt

           Control Systems Specialist
           Messier-Dowty Inc.
           574 Monarch Ave.
           Ajax, Ontario
           L1S 2G8
           Canada
           Tel: (905) 683-3100
           Fax: (905) 683-6983
           E-mail: Kyle.Schmidt@messier-dowty.on.ca

Automating the San Francisco Bay Model with LabVIEW

Dave Weisberg
Principal
Cal-Bay Systems, Inc.

Products Used. FieldPoint, LabVIEW, PID Control Toolset, SQL ToolKit, ComponentWorks™, Motion Control, signal conditioning (SCXI), and data acquisition (DAQ).

The Challenge. Automating one of the only physical hydraulic models in the U.S. to simulate precise tidal conditions and acquire large amounts of data.

The Solution. Providing a cost-effective, cutting-edge, fully networked, closed loop system with full flexibility and expandability for the future with Cal-Bay Systems, Inc. and National Instruments products.

Introduction

The San Francisco Bay model (Figure 5-23) was originally designed to test the Reber plan, which was a radical plan in the late 1950s to provide fresh drinking water for the ever-expanding San Francisco Bay Area. The plan involved damning off the northern and southern sections of the San Francisco Bay and using it as a fresh water reservoir. More recently, the model has been used to measure the environmental effects of deepening shipping channels through the Concord straits to allow oil freighters access to expanding refineries. Particular attention is paid to water levels, the mixing of fresh and salt water, and current flows through the narrows of the estuary.

The original data acquisition in the model was done manually. Operators would visually inspect the instruments and log the data manually into a logbook. Tides were controlled manually using computer-aided manufacturing (CAM) systems that controlled inflow and outflow valves to allow the cor-

Figure 5–23
The San Francisco Bay model

rect amount of water to be introduced into the model. Manual adjustments were made if the tide drifted from the desired signal.

In 1982 an HP1000/2250 computer system was installed to automate data acquisition. Reel-to-reel tapes were used to archive data from tests that were run in the model. The tide control was partially automated using the HP1000 but, due to memory limitations of 64K, full control was not feasible. Periodic manual adjustments were still required.

System Upgrade

The need for expanding the channel count, implementing fully automated tide control, and presenting data to the entire Army Corps of Engineers via the Internet led to the decision to replace the old system with a cost effective

PC-based system. The PC was outfitted with a wide range of automation hardware from National Instruments. LabVIEW software was selected as the programming tool of choice. Cal-Bay Systems, Inc. of San Rafael, California, was contracted to provide systems integration and software programming services to implement the new system.

Automation Requirements

There were three areas of automation and integration to tackle for the upgrade:

- **Tidal control** — Precise tidal conditions at the Golden Gate Bridge must be reproduced in the model to within .01 inches of corresponding actual data collected by the Army Corps of Engineers for that location. This data includes full moon (spring tides) and no moon (neap tides) with tidal variations between high and low tide of up to 7 feet (.84 inches in the model).

- **High-channel count data acquisition** — Over 150 sensors are located around the model to acquire information. Data must be collected and archived every two seconds. All sensors must be calibrated before every test. And all data must be available on the Internet so it is readily accessible from all offices of the Army Corps of Engineers. The Corps of Engineers must be able to view and compare data over time, across different tests, and at different locations.

- **Distributed control** — Flow controllers are used to simulate river flows from the Sacramento Delta and are located at several distant places around the model. These needed to be controlled from a single central source.

System Configuration

The Tide Control system is comprised of a 75 horsepower, three-phase pump that is used to drive salt water into the model at a constant flow rate. Three butterfly valves are positioned in the Pacific Ocean area of the model. Each value is controlled by a servo motor and allows water to flow out of the model at a varying rate. The outflow is gravity fed. When the valves are

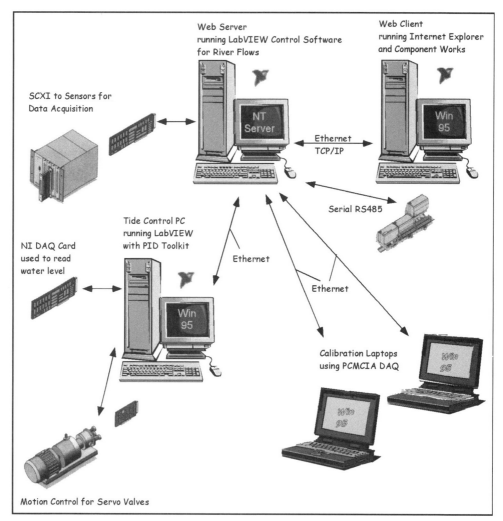

Figure 5-24
System configuration

closed, the model fills with water (high tide). When the valves are open, the model empties (low tide).

Three ultrasonic level sensors, placed in different physical locations, are used to monitor the water level in the model. The actual water level is derived by averaging the three, therefore filtering out any oscillations caused by high frequency ripples or waves. A National Instruments DAQCard, LabVIEW, and the LabVIEW PID Control Toolset are used to monitor and filter the water level data and determine the proper outflow for the desired tide. Using National Instruments motion control technology, LabVIEW converts the desired outflow to motor positions on the servo-controlled butterfly valves. A setpoint profile, representing 3.6 years of tide level data, is fed into the PID controller as the desired tide signal.

Tuning the tide control system was the most difficult part of the application. Since tide control had never been accomplished as a fully automated system, there was a lot to learn. Software techniques were implemented by Cal-Bay Systems that allowed a further understanding of the dynamics of the hydraulic system. Frequency analysis was done to determine the harmonics of the oscillations, and the data was used to help tune the PID tide control algorithm.

The data acquisition portion of the application consists of a National Instruments SCXI chassis housing eight SCXI 1100, 32-channel multiplexer modules. This allows a total of 256 channels of analog input, leaving plenty of room for future expansion. Using LabVIEW, Cal-Bay Systems implemented a data acquisition program that scans all 256 channels at a rate of once every two seconds. LabVIEW collects the data and scales it appropriately using third order calibration equations that are generated by the calibration software. The data is stored in a Microsoft Access database, and an Internet application using Microsoft's Internet Information Server was written to serve up the data on the Internet for viewing and analyzing in a Web browser. ComponentWorks Active X technology from National Instruments is used as part of the Internet application for analysis of the data and presentation in the Web browser.

Calibration of the sensors used to collect salinity and velocity information is critical before every test is run. Because of their portability, laptops are used around the model to perform the calibrations and send calibration information back to the server via wireless Ethernet. Cal-Bay authored a LabVIEW application that interfaces to National Instruments PC DAQ Cards for data collection on the laptops and relays the calibration information back to the main data acquisition server.

The simulated river flows are located in distant locations around the model, making them an excellent candidate for National Instruments FieldPoint modules. Analog input modules monitor the flow rates. The data is fed through multiple PID loops using the PID Control Toolset and LabVIEW. Analog output modules vary the flow controllers to achieve the desired flow rates. The entire system is integrated via a local area network to allow each different component to communicate its status and information. The entire system configuration is illustrated in Figure 8-24.

Results

The San Francisco Bay Model came on line in the third quarter of 1998. One of the first studies using the new system was conducted to determine the need for rebuilding the Sacramento Levies that were destroyed by El Nino during the winter of 1997–1998.

■ Contact Information

Dave Weisberg

Cal-Bay Systems
3070 Kerner Blvd.
San Rafael, CA 94901
Tel: (415) 258-9400
Fax: (415) 258-9288
E-mail: Dave@calbay.com

Virtual Balancing Equipment for Rigid Rotors

Ricardo Jaramillo
Manager
Ricardo Jaramillo y Cía

Daniel Jaramillo
Engineering Assistant
Ricardo Jaramillo y Cía

Products Used. LabVIEW, data acquisition card.

The Challenge. Building a cost-effective and easy to operate balancing system incorporates all the calculations needed for shop and field balancing of rigid rotors.

The Solution. Creating a system running National Instruments LabVIEW under Microsoft Windows that reads two accelerometers and a photocell and performs the calculations needed for two plane dynamic balancing, plus many additional calculations.

Introduction

There are many electronic equipment manufacturers that make balancers for rigid rotors, but these balancers do not allow the user to make all the calculations needed during the process. In addition, they do not allow the user to save information in order to average reports later on.

The virtual balancer described here is an instrument that is easy to use that can conduct many tests. It can acquire signals from two accelerometers and a photocell, calculate the balancing weights, simulate the dynamic response of the machine, plot signals and its spectra, plot the vector diagrams, and translate the calculated weights to real world corrections. The system can also calculate angular, axial, and radial translations of weights and store the influence coefficients and rotor characteristics for future balances of the same

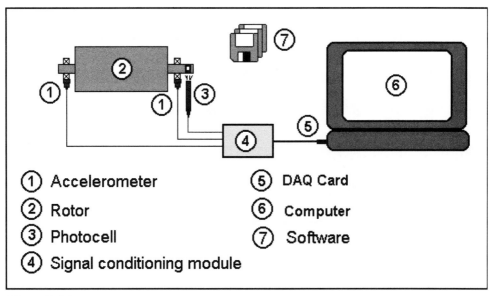

Figure 5-25
Hardware configuration of the rigid rotor balancer

rotor geometry. It can do all of this and be produced at a fraction of the cost of electronic balancers in the market.

System Hardware Configuration

The hardware is composed of the following elements (see Figure 5-25):

- **Accelerometers** — There are two of them, and they must have built-in amplifiers and a sensitivity in the middle range (about 100 mV/g).
- **Photocell** — It works at short distances (less than one inch) and has very low battery consumption.
- **Signal conditioning module** — A simple electronic circuit conditions the signals from the two accelerometers and the photocell. It can work from batteries or from 110V AC.
- **DAQ card** — The DAQ card used depends on whether the computer is a desktop or laptop. The minimum requirements are: 50 kHz sampling rate, at least three channels, and 12 bits of resolution.

National Instruments has a wide variety of cards that meet these requirements.

- **Computer**—The computer must run Microsoft Windows 98 or 95. If the system is going to be used for field balancing, a laptop computer must be used.

Virtual Balancer Options

The main panel asks the user to select one of the following options for balancers and reports:

- 1-plane, one for accelerometer (static balancer)
- 2-plane, 1 accelerometer (dynamic balancer)
- 2-plane, 2 accelerometers (dynamic balancer)
- A simplified summary report
- A detailed report that prints one balancing job per page.

The reason for having a 2-plane, 1 accelerometer instrument is because the user can proceed with the balancing work even if one of the accelerometers is damaged. In this case, the user would take one reading, move the accelerometer to the other measuring plane, and take a second reading.

Two-Plane, Two Accelerometers Balancer

Figure 5-26 shows the front panel of the 2-plane, 2 accelerometers virtual balancer. From this panel a user can:

- View the signal from the accelerometers and the photocell
- View the amplitude spectrum and the phase of the signal at 1 x RPM corresponding to the accelerometers
- Proceed with the acquisition
- Observe the trigger level for the photocell signal, which is calculated by the program

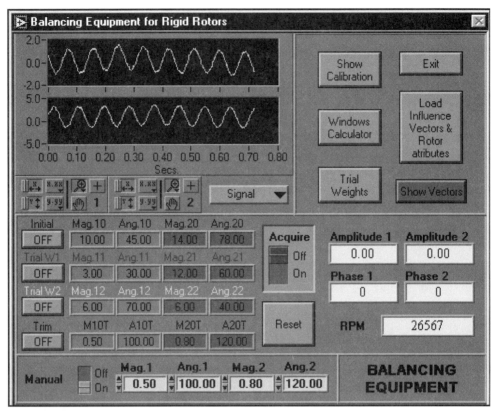

Figure 5-26
Front panel of the 2-plane, 2 accelerometers rigid rotor balancer

- Pass the amplitude and phase readings to the module that performs the balancing calculations
- Call the calibration window in order to see or to change the calibration of the accelerometers; call the calculating module or the trial weights window
- Load influence coefficients used in the past
- Simulate two-plane initial and trial conditions in order to study the behavior of the rotors
- Reset the instrument.

Chapter 5 • General Test

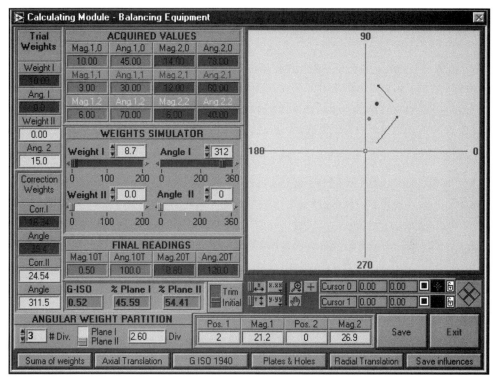

Figure 5-27
Front panel of the calculating module

First the user takes three readings—the initial reading, a reading with one trial mass on plane 1, and a reading with a trial mass on plane 2. Then, the user calls the calculating module (see Figure 5-27).

In this module, the user can:

- View vector diagrams that represent the initial unbalance of the rotor and its behavior with the trial masses. The colors of the points and lines in the graph are coordinated with the colors of the indicators and controls in this window as well as others throughout the program.

- View the values of the trial masses and their positions, the values of the calculated correction masses and their positions, and the magnitude and phase readings for the initial, trial, and final states.

- Simulate correction masses with angular positions and to see how the rotor behaves if unbalance.

- View a table of balancing quality grades according to ISO-1940.
- Call different windows for special calculations dealing with balancing such as vector addition of weights; dimensioning of holes, plates, and electrodes; axial movement of correction weights; and radial movement of correction weights.
- Switch the vector diagrams of initial and final conditions back and forth and to see the initial and final balance quality after the correction weights have being fixed to the rotor.
- Save the influence coefficients and rotor characteristics for future balancings of the same type of rotor (identical geometry).
- Save all relevant information in order to produce reports later.
- View the angular correction weight partition for both planes and the obtained balance quality grade according to ISO.

Innovative Features

The first innovation in our balancer is its virtual nature—we think it is the first virtual dynamic balancer ever built. Other innovations deal with the calculations available to the user. The correction weights simulator is useful when the behavior of the rotor is not linear due to external conditions such as excessive clearances in rotor bearings, resonance, and excessive deformation due to unbalance. This instrument also displays calculations for the axial, radial and angular translation of correction weights and converts the correction weights into equivalent holes, plates, or electrodes for the most common fabrication materials used in industry. Another innovation is the rotor response simulator that allows the user to calculate the correction weights for any set of initial and trial mass conditions.

Auxiliary Windows

From the acquisition and calculating windows users can view other windows that allow them to perform calculations dealing with the balancing process. Some of these are:

- **Trial weights window** — This window lets the user calculate a trial weight and to enter information that the software uses to calculate the correction weights and the balance quality grade.

- **Axial translation of weights window** (see Figure 5-28) — This window allows the user calculate the axial translation of correction weights.

- **Balance quality grades window** — This is a table with the balance quality grades of most common machinery according to ISO-1940.

- **Plates and holes window** — This window is used to translate the correction weights in grams or ounces to equivalent plates, holes, blind holes, and electrodes of different materials and dimensions. The user can work in grams or ounces or in millimeters or inches. It takes into account the angular spread of holes (if the user specifies several) in its calculation. It also warns the user if there are too many holes (more than 180° spread).

Advantages of the Equipment

There are five main advantages of this equipment when we compare it with the electronic balancers available on the market. These are:

- **Saves time** — Most of the calculations available on this system are not included in other balancers.

- **Cost-effective** — The total cost of this equipment (including the computer) is about 1/3 of the average cost of other balancers.

- **Complete** — All of the calculations needed to perform balancing work are included in this system.

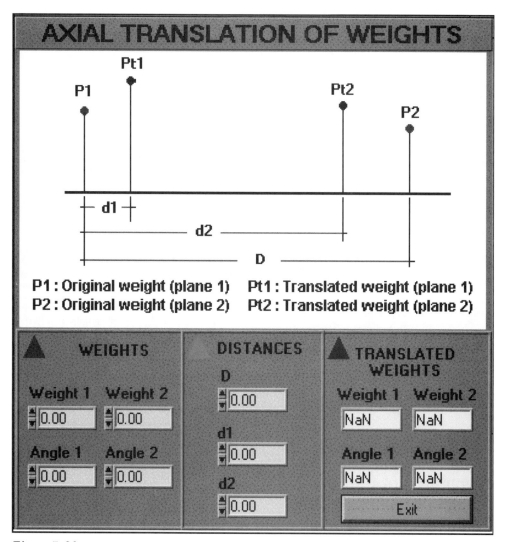

Figure 5–28
Calculator for axial translation of weights

- **Upgradable** — This balancer, as with any other virtual instrument, is very flexible and easy to upgrade compared with stand-alone instrument.

- **Added Functionality** — Besides having the best balancing equipment your money can buy, the system also includes fully functional personal computer.

Results

This virtual equipment has many utilities not available in regular electronic balancing equipment but that are easy to include in a virtual instrument. It is very hard and costly to generate these with electronic hardware.

Virtual instrumentation is very successful because it is standing on the shoulders of a giant — the personal computer. This balancer is not an exception to that rule. None of the manufacturers of electronic balancers can compete in technology with the computer manufacturers, and the reason for that is the size of the PC market, compared to the market for electronic balancers.

Contact Information

Ricardo Jaramillo

> Manager
> Ricardo Jaramillo y Cia
> Crr. 32 No. 10-121
> Medellin, Colombia
> Tel: (574) 266-5469
> Fax: (574) 266-5469
> E-mail: rjaramillo@epm.net.co

Daniel Jaramillo

> Engineering Assistant
> Ricardo Jaramillo y Cia
> Crr. 32 No. 10-121
> Medellin, Colombia
> Tel: (574) 266-5469
> Fax: (574) 266-5469
> E-mail: djaramil@delta.eafit.edu.co

LabVIEW-Based Automation of a Direct-Write Laser Beam Micromachining System

Ampere Tseng
Professor
Arizona State University

George Vakanas
Research Assistant
Arizona State University

Bill Watson
Test/Design Engineer
BF Goodrich Aerospace Integrated Systems

Products Used. National Instruments LabVIEW, Spectra Physics CIM-2 interface (for the GCR130-10 series lasers), Parker/Compumotor AT6400 code (for XYZ stepper-motorized-stage), and Matrox Imaging software (for CCD PCI video card interface).

The Challenge. Integrating the controls of three components of a laser-beam micro-machining system for process monitoring and automation using an intuitive graphical instrumentation environment such as LabVIEW, increasing productivity in experimental research, and adding value in university-level laboratory training.

The Solution. Programming an automation software using the capabilities of LabVIEW for external function calls to third-party DLL files to communicate instructions via the Spectra Physics CIM-2 laser interface, the Compumotor AT6400 motorized XYZ stage indexer, and the Matrox video interface card.

Introduction

Our project involves the set-up and automation of a direct-write laser beam micromachining system to meet the needs of mechanical engineering graduate-level research and laboratory training in advanced manufacturing processes at Arizona State University (ASU). The laser-materials processing, as applied to the mesoscopic and microscopic scales, is a nonconventional, yet fundamental manufacturing/fabrication process with a number of industrial applications:

- Drilling of turbine blades
- Drilling of inkjet micronozzles
- Scribing, marking, and cleaning of semiconductor or microelectronic products
- Restructuring (cutting and linking) aluminum and copper interconnects on microelectronic devices and application-specific integrated circuits (ASIC)
- Recording, reading, and erasing of information on magneto-optical media
- High-aspect ratio drilling of biomedical devices (e.g., catheters)
- Etching and scribing microchannels on wafers for electronic cooling and thermo-fluidic MEMS applications.

The basic direct-write laser machining system consists of the laser source, the beam guiding/focusing optics, a charge coupled device (CCD)-based imaging system, and a step-motorized XYZ stage. The above components rest on a TMC vibration-free, optical table, providing 8-ft by 5-ft of working space for beam guidance and focusing the laser beam on the target. The high-power laser flux attained at the focal point of the optical system is able to remove (ablate) material from the workpiece in a precise and repeatable manner. The step-motorized XYZ stage is used to trace the motion of the desired CAD pattern on the target workpiece. The stage has a 6-in x 6-in x 4-in travel in the x, y, z directions respectively. The CCD-based imaging system is used to record images of the etched patterns on the workpiece targets at variable magnifications (300 to 2000X).

The components of the experimental set-up that have been successfully implemented and computer-controlled are:

1. The 0.5-micron precision step-motorized XYZ stage, controlled via the AT6400 code, which provides instructions to the three lead-screw stepper-motor controllers via an indexer.
2. The 450mJ pulsed Nd:YAG laser (with pulses of 2 ns, 8 ns and 125 μs) operating at the frequency-doubled second harmonic (green, 532 nm) wavelength.

3. The high-resolution ½-in CCD-based imaging system, controlled via the Matrox interface.

The Micromachining Application: Process Automation

The experimental research program aspires to establish process windows for Nd:YAG laser-based machining on a diversity of materials in thin or thick films and sheets. Material processing operations of interest include scribing, high-aspect drilling, and fine-line cutting based on customer driven designs and specifications. The research program currently experiments with metals (aluminum, steel, titanium) and metal thick films on silicon wafer substrates (copper and permalloy). Student and researcher perceptions are realized with a CAD tool of their choice and saved or translated in a standard engineering drawing format (i.e., DXF file, HP-GL plot file or G-code NC program). Although commercially available, the code to directly translate the CAD code to the AT6400 code to drive the XYZ-stage is not in place yet. Instead, the CAD file is recast into the AT6400 code in a raster format, which may not be the optimal way for path tracing. At the request of LabVIEW, the pattern code is executed by the motorized stage to realize the relative motion between the laser and the workpiece. LabVIEW also turns the laser on and off and adjusts its laser power and pulse rate. The process is concurrently monitored by a CCD-based imaging system.

Sample CAD designs such as simple microchannels formed by straight line segments and arcs, the ASU text pattern (Figure 5–29), and gear designs are routinely scribed, drilled, and cut on target workpieces.

System Requirements

System automation relies on an Intel Pentium II PC with on-chip math support, running at 332 MHz, 64 MB of RAM, 7 GB of hard disk storage and a 32-bit color ATI 3D RAGE PRO video card and monitor. The computer is connected with the laser through the laser CIM-2 programmable interface module via the serial port. The computer is further connected to the XYZ-stage indexer and the CCD camera system via PCI interfaces. All third-party systems that are in use support GPIB communication and plans for upgrades exist when its use is justified.

Figure 5-29
The ASU text pattern, scribed on aluminum 6061. The path was coded in AT6400 format and downloaded to the XYZ stage indexer via LabVIEW. Simultaneously, the Nd:YAG laser pulsing is controlled. Laser focus spot ~30 microns, energy: 200 mJ/8 ns

The software is programmed under the Microsoft Windows 95 platform. Requirements such as the need for an intuitive, graphical, dynamic, and self-documenting user interface and the ever-expanding research needs for increased project complexity favored LabVIEW as the programming environment of choice for the ASU laser machining application. The software project development cycle for the first functional module for concurrently controlling two of the three components was completed in 5 weeks (Figure 5-30).

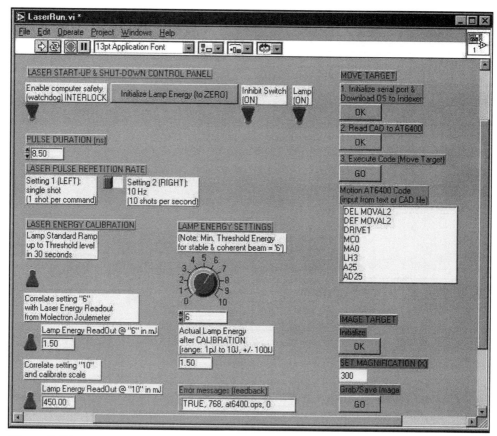

Figure 5-30
The LabVIEW master control panel of the ASU laser micromachining system

Database Concepts: Expanding the Materials Space

Due to the great number of distinct physical and chemical phenomena involved in the laser-material interaction, a general laser-material interaction theory is not yet available. Depending on the incident radiation and material properties, such microscopic phenomena usually include: radiation absorption and reflection by the surface and/or by the bulk material, excitation of electrons and relaxation into phonons (and subsequent heat), and plasma coupling and explosive effects. In macroscopic phenomenological terms, the

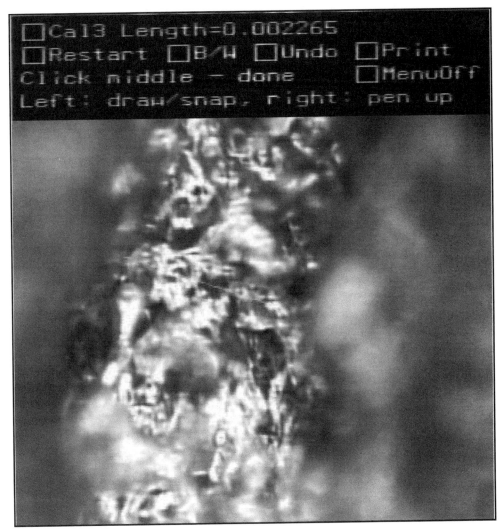

Figure 5–31
A laser cut of ~50 mm on a 4x4" aluminum 6061 plate (focus on the bottom of the trough). Laser power: 200 mJ, wavelength=532 nm, pulse: 8 ns, 10 Hz. Notice the adverse heat-affected zone (HAZ) that results in undesirable surface roughness.

operative physical phenomenon in laser-material interaction may be characterized as photomechanical, photochemical, or photothermal (evaporation, melting, heat treatment). For an example of adverse melting and the so-called heat affected zone (HAZ) see Figure 5–31.

Establishing a laser process based on new material is not as trivial as it sounds. In order to make the laser-materials processing systems agile with respect to different materials, a great deal of work is needed due to the diversity of phenomena operative for each type of material. Thus, some kind of classification would indeed be a great aid in deciding upon criteria that could define process windows given the material to be processed, the laser source characteristics and the CAD geometry to be machined.

A comprehensive list of process parameters for laser direct writing is shown below:

- Laser power/energy (pulse energy E_p)
- Wavelength (λ)
- Laser beam size $w(z)$ and focus spot size (radius R)
- Thermal diffusivity (α)
- Material properties, reflectivity, and melting and transition temperatures
- Composition, impurities, and physical microstructure
- Coherence effects
- Nonuniform heating
- Instabilities
- Interface/adhesive characteristics, thermal mismatch, and differential thermal expansion
- Motorized stage scanning speed (u)
- Material removal rates.

Designing experiments with the above parameters as variables is not efficient and probably a formidable task. Therefore, nondimensional groups are sought. For a simple physical process, nondimensional parameters can actually be generated by a normalization procedure of the governing equations (e.g., continuity, momentum, energy, Maxwell's). For more complex phenomena, empirical nondimensional parameters, hinted by either theory or experiments, are used. The nondimensional parameters, shown in Table 5-2, are used as an efficient tool to provide useful criteria for the operative laser-material phenomenon at hand. Material properties are first hardcoded using the database concepts of lists and look-up tables in a spreadsheet or text/ASCII format. With these material properties, input laser characteristics, and desired simple CAD features, the three criteria are calculated and our LabVIEW virtual instruments (VIs) that we develop in house can use them to estimate and propose usable process windows.

Table 5-2 Relevant Non-dimensional Parameters in Laser-materials Processing

Non-dimensional Parameter Name	Expression	Physical Significance	References
Diffusion vs. absorption length	$L^* = \dfrac{\sqrt{4\alpha\tau}}{\dfrac{\lambda}{4\pi k}}$	Criterion for surface vs. bulk absorption in 1-D conduction	Grigoropoulos C.P. (1998), Laser, Optics & thermal considerations in ablation experiments, *Exp. Meth. In the Phys. Sciences*, Vol. 30, Academic Press.
Diffusion speed vs scanning speed	$U^* = \dfrac{uR}{\alpha}$	Criterion for surface vs bulk absorption in convection (relative motion)	Abakian and Modest (1988), Evaporative Cutting of a Semitransparent Body With a Moving CW laser, Trans. Of the ASME J. of Heat Transfer, Vol. 110, p.924-930.
Absorption-splashing parameter	$\beta^* = \dfrac{t_p}{t_r} = \dfrac{4kE_p}{\lambda R^2 \rho H_{ev}}$	Criterion for splashing (=subsurface boiling)	Ready (1963), Effects of high-power radiation Schwartz, H. and H.A. Tourtellotte (1969). J. Vac. Sci. Technol., Vol. 6(3), p. 373-378.

Nomenclature

τ = diffusion time
t_p = pulse duration
t_r = relaxation time ~1% of Debye time
k = extinction coefficient = imaginary part of refractive index

ρ = substrate density
H_{ev} = heat of evaporation [J/kg]

Results

At the current level of development, LabVIEW was used and proved useful for implementing process control and automation. The advantages of the LabVIEW choice, in the current application, could be summarized as:

- Compatibility and universal communication with a diversity of third-party proprietary computer codes (Parker/Compumotor AT6400, Spectra Physics CIM-2 laser interface, Matrox video interface)
- Modularity, which leads to code reusability, ideal for time-consuming, complex and evolving projects
- User-friendliness and self-documenting capability, which proves extremely useful for online tutorials.

We look forward to upgrading the ASU LabVIEW package to include measurement and analysis of workpiece parameters of interest such as: surface roughness, hole edge profile, and hole taper and tolerances of the resulting machined geometries.

Future Improvements

Given the possibilities of LabVIEW along with an abundance of automation and measurement that needs to be performed, the number of future improvements are virtually endless and only limited by budget, system complexity, and imagination. Given the envisioned use of the system in research and education, the following is a nonexhaustive list of our planned future improvements:

1. *In situ* laser power monitoring and implementation as an additional LabVIEW-controlled component of the system. This will allow validation and precise measurement of the laser power level used in each experimental run. The hardware, MOLECTRON's Joulemeter, is already available and provides direct analog and digital readout with RS232 capability. An additional optical component (beam splitter) is needed along with an additional VI module programming.
2. *In situ* vibration monitoring and implementation as an additional LabVIEW-controlled component of the system. Vibration is indeed a

major issue in precision machining, where positioning and relative motion between the tool and the workpiece should be strictly bounded

3. Offline surface roughness characterization of the machined, etched, or scribed feature

Conclusions

LabVIEW proved an extremely efficient and highly operational automation solution for the laser micromachining project. Overall, the ASU LabVIEW VIs developed in-house are a significant contribution in the field. They greatly reduce the number of calibration experiments, thus increasing performance and productivity in meaningful experimental work and in industrial settings of laser materials processing at the macro-, meso- and micro-scales. Also, with the relevant LabVIEW VI libraries that resulted out of this project, new automation tools have been put in place and will also provide a good software module library for our next round of LabVIEW-based software development. LabVIEW has become an integral part of our laboratory's vision for establishing and maintaining an agile, automated laser-based micromachining process.

Acknowledgements

Credits for this research are due to Scot Robertson of SERTEC for his assistance with the laser microscope optical system, Ron Swift of MITSUBISHI SILICON AMERICA for providing us with material samples, and Tim Karcher of the ASU Center for Solid-State Sciences (CSSS) for the preparation of sputtered metal thick films.

■ Contact Information

George Vakanas
  Research Assistant
  Manufacturing Institute, MC5106
  Department of Mechanical Engineering
  Arizona State University
  Tempe AZ 85287-5106
  Tel: (480) 965-9168
  Fax: (480) 965-2910
  E-mail: george.vakanas@asu.edu

Index

A

acoustic quantification, 78
ActiveX, 162
actuator endurance tests, 214
algorithm, 156
anechoic chamber, 143
angle of incidence, 104, 109
angular scanning ellipsometer, 104
angular scanning optics, 106
antennas, 142
anti-aliasing filters, 93
assembly language, 133
asynchronous machine, 180
audio test, 43
automotive communication bus, 60
automotive emission control, 29

B

back-scattered light, 203
baseline drift, 80
beam profiler, 205
blood pressures, 92
Butterworth filters, 83

C

C structure, 148
C/C++ language, 133, 162
calibrations, 221
camera, 202
CANDU, 186
cardiac cycle, 80

cardiac output, 92
cardiovascular, 76
cathode ray tube (CRT), 152, 195
charge density, 114
charge-coupled device, 45, 87, 121, 196
Chebyshev filters, 83
Classical Electron Oscillator model, 111, 118
coherence effects, 238
collimating optics, 106
commodity off-the-shelf, 208
computer control system, 106
concurrent execution, 201
context switching, 205
control, 173
controller area network (CAN), 45
counters, 127
Crash Avoidance Metrics Partnership (CAMP), 36
cross correlation, 73

D

data acquisition, 71, 104, 106, 127, 133
data analysis, 109
datalogging, 84, 126
DC machine, 176
DC/DC converter, 177
digital image flaw detection system, 195
dipolar oscillations, 117, 119
distal pressure measurements, 81
Distributed Control, 219
dual-processor, 201
dynamic link library (DLL), 72
dynamometer, 22, 25, 27

243

E

ECG, 74
echocardiography, 77
ejection fraction, 77
electric dipole, 112
electric vehicle, 5
electrocardiogram (ECG), 70
electrohydraulic servo valve, 211
Electron Beam Physical Vapor Deposition
    (EBPVD), 121
electron beams, 122, 123
electron behavior, 111
ellipsometric parameters, 108
ellipsometry, 103
end-diastole, 77
End-Diastolic Pressure-Dimension Ratios
    (EDPDR), 83
endoscopy, 89
end-systole, 77
End-Systolic Pressure-Volume Relationship
    (ESPVR), 82
endurance testing, 208
energy calculations, 127
Ethernet, 55, 130
Excel, 95

F

Fast Fourier Transform (FFT), 108
filtering, 185
flow controllers, 219
flutter, 43, 50
FM Frequency Response test, 63
focusing optics, 105
FORTRAN, 71
forward collision warning (FCW), 35
frequency analysis, 221
fuel injection rate control, 31

G

Good Laboratory Practice (GLP), 92
GPIB, 163
GPS, 35
GRASS instruments, 72

H

Hall-effect, 177
Hamiltonian, 112
heart period variability (HPV), 70
heart rate, 92
heat affected zone (HAZ), 237
heavy water detritiation, 186
high voltages, 177
High-Channel Count Data Acquisition, 219
HiQ, 95
hydraulic actuators, 208
hydrogenic atoms, 114

I

image processing, 121
imaging system, 233
instrument control, 181
interactive phase-delays, 81
Interchangeable Virtual Instruments, 163
Internet, 173, 181, 182, 218
Iridium system, 147
isolation amplifiers, 177
isotopic exchange, 185

J

J1850, 45
JPEG, 57

L

laser, 203
laser beam, 200
laser diode, 203
laser power monitoring, 240
laser-material interaction, 236
latency, 70
load cell, 211
local area network, 35

M

Matlab, 108
Matlab/Simulink, 174
mechanical engineering, 232
Microsoft Access database, 46
motion control, 104, 106
motion control system, 106
multipath distortion, 143
multitasking, 201
multithreading, 201
myocardium, 76

N

needle-detection algorithm, 12
noise immunity, 177
nondestructive test system, 198

O

odometers, 19
operating system, 205
optical character recognition (OCR), 20
optical power, 107

P

parallel processing, 201
partially reflective optic, 202
patient monitoring, 92
photochemical, 237
photodiode, 203
photolithography, 126
photomechanical, 237
photo-thermal, 237
physiological preamplifier, 72
physiological signals, 76
precision machining, 240
pressure curves, 192
pressure-area analyses, 77
pre-trigger, 203
process monitoring, 232
programmable logic controller (PLC), 129
progressive scan mode, 203
proportional integral derivative (PID), 209
pulmonary function, 92
Pulse Width Modulated (PWM), 45

Q

QRS complex, 70
quantum mechanics, 111, 117

R

radial wave function, 114
radiation pattern, 142
radio frequency (RF), 35, 60
real-time, 93, 127
remote control, 132, 152, 153
remote device access (RDA), 57
retarder, 108
root-mean-square (RMS), 48

S

sensitivity, 43
serial, 163
servo motor, 219
Shrödinger's equation, 112
signal amplification, 185
signal to noise ratio (SNR), 43
simulation, 173
sleep disorder, 70
slip-ring induction motor, 176
spatial color distributions, 88
spectral color distributions, 88
spectrum analyzer, 152
speedometers, 11, 19
spherical harmonic, 114
Statistical Process Control, 12
Stokes parameters, 108
stroke work, 81
Structured Query Language (SQL), 43, 192
synchronous machine, 176

T

tachometers, 19
tank cannons, 192
TCP/IP, 39, 130
temperature, 121
test audio, 43
test executive, 162
thermal barrier coatings, 125
thermal cycling, 6
thermal diffusivity, 238
thermocouple relay module, 27
thermocouples, 122, 123, 186
three-phase voltages, 180
Tidal Control, 219
tide control, 218
time-dependant perturbation theory, 111, 113
timeslicing, 204
time-varying elastance, 81
tire tread loading, 23

U

ultrasonic level sensors, 220
ultrasound machines, 76
ultrasound signals, 78

V

vacuummeter, 131, 133
ventricular volume, 77
vibration monitoring, 240
vibration tables, 53
VIs, 210
VISA, 154
Visual Basic, 174
Visual C++, 174
visualization, 173
voltage dividers, 177
voltage input module, 27
voltage output module, 27
voltage transducers, 177

W

wave function, 112
waveform generator, 156
wavelength, 238
wavelength division multiplexing (WDM), 152
Weibel radar unit, 193
wireless, 35
wireless LAN, 37
wound healing, 87
wow, 43, 50

X

X-ray, 126, 197
X-ray digitizer, 127
X-ray film photographic inspection system, 195
X-ray industrial television, 195

Keep Up-to-Date with
PH PTR Online!

We strive to stay on the cutting-edge of what's happening in professional computer science and engineering. Here's a bit of what you'll find when you stop by **www.phptr.com**:

- **Special interest areas** offering our latest books, book series, software, features of the month, related links and other useful information to help you get the job done.

- **Deals, deals, deals!** Come to our promotions section for the latest bargains offered to you exclusively from our retailers.

- **Need to find a bookstore?** Chances are, there's a bookseller near you that carries a broad selection of PTR titles. Locate a Magnet bookstore near you at www.phptr.com.

- **What's New at PH PTR?** We don't just publish books for the professional community, we're a part of it. Check out our convention schedule, join an author chat, get the latest reviews and press releases on topics of interest to you.

- **Subscribe Today!** **Join PH PTR's monthly email newsletter!**

   Want to be kept up-to-date on your area of interest? Choose a targeted category on our website, and we'll keep you informed of the latest PH PTR products, author events, reviews and conferences in your interest area.

   Visit our mailroom to subscribe today! **http://www.phptr.com/mail_lists**